The World's Worst Problems

Walter Dodds

The World's Worst Problems

 Springer

Walter Dodds
Division of Biology
Kansas State University
Manhattan, KS, USA

ISBN 978-3-030-30409-6   ISBN 978-3-030-30410-2   (eBook)
https://doi.org/10.1007/978-3-030-30410-2

This Springer imprint is published by the registered company Springer Nature Switzerland AG
The registered company address is: Gewerbestrasse 11, 6330 Cham, Switzerland

Acknowledgments

I thank Dolly Gudder, Anne Schechner, Lindsey Bruckerhoff, Sky Hedden, Crosby Hedden, Garrett Hopper, James Guinnip, Liz Renner, Casey Pennock, and Keith Gido for providing input on the book. Sherestha Saini kindly and patiently considered the book as Editor at Springer. I especially thank all the students and others who took the time to discuss many of the ideas in this book with me so thoughtfully. Also, I am grateful for my seminar hosts over the years who have indulged me and allowed me to speak on this topic including Davi Gasparini Fernandes Cunha and Antoine Leduc. I thank Kansas State University for support during while preparing this book.

Contents

About the Author

Walter Dodds is a University Distinguished Professor at Kansas State University and an Edwin G. and Lillian J. Brychta Chair in Biology. He specializes in environmental aspects of freshwater ecology. He has numerous peer-reviewed publications and has authored several books including *Humanity's Footprint: Momentum, Impact, and Our Global Environment* (on global environmental issues) and *Laws, Theories, and Patterns in Ecology* (a graduate overview of ecology). He is a Fellow of the American Association for Advancement of Science, the Society for Freshwater Science, and the Association for the Sciences of Limnology and Oceanography.

Chapter 1
Introduction

This woodcut from Eloy D'Amerval's 1508 Livre de la Deablerie purports to show Eloy at work. In fact, it was a woodcut used by early French printers to represent other authors as well. (Public Domain Image)

It is not easy to think rationally about how humans treat each other and the environment upon which they depend. If you are not suffering, are you to blame for those who are? I personally have skirted serious consideration of parts of these questions for quite some time because of my fears, my avoidance of guilt, and my being overwhelmed by the enormity of the issues. Still, I will attempt to approach these hard questions bringing as many facts and as much logic as possible to bear on the issues. The response of the reader can be kill the messenger, ignore the facts, or consider what humanity can do about these problems and how their actions can make a positive difference. I am optimistic that the reader will chose the third of these options.

© Springer Nature Switzerland AG 2019
W. Dodds, *The World's Worst Problems*,
https://doi.org/10.1007/978-3-030-30410-2_1

People have an incredible will to survive and are extraordinarily adaptable and intelligent. These abilities make it possible that humanity could survive almost anything. Our species has weathered warm periods and global glaciations. Huge shifts in climate led to much drier or wetter conditions, requiring dramatic shifts in the behaviors of primitive people. Somehow, some of them learned how to cope and multiply in a changing world.

Our genus evolved 2.3–2.4 million years ago on the relatively warm African continent. Our ancestors inhabited a world with numerous large and scary predators. Our predecessors had small teeth, modest physical strength relative to other animals, and eventually a few puny weapons for defense; disease and starvation were often just around the corner. Life expectancy was short. Yet, *Homo sapiens* thrived and spread to every corner of the planet, from the driest of deserts to the coldest polar habitats. We adapted our behaviors and social systems to overcome and become the most dominant species on Earth.

We are intelligent and flexible; we develop and transmit information through cultural evolution. We control our local environment to feed and shelter ourselves. These abilities have also ultimately led to the possibility that we could destroy ourselves; now we have the capacity to alter our entire biosphere. Our behaviors, technology, and lifestyles lead to global problems because we are the most dominant organism in Earth's history. We control so much energy and material power that our actions influence the air we breathe, the water we drink, and every other plant and animal on Earth.

Yet, we cannot escape the physical and biological realities of life that constrain our existence. We need air to breathe, food to eat, shelter from the elements, and some degree of security. We can survive under horrible conditions and extreme suffering, yet such survival is only just that. Many have come to expect a quality of life better than one that is just long enough for reproduction and propagation of the species. Bad teeth, poor health, stunted growth from lack of nutrition, and disease no longer need to be common conditions for people. Proportionally fewer people suffer now than ever before in history; health is better, food more reliably present, lifespan longer, more diseases can be treated, and many live in greater comfort.

The unprecedented and widespread high standard of living that people in developed countries have come to enjoy has set a very high bar for the rest of the world to attain. The lifestyle of the privileged also is testament to the fact that a life only modestly affected by hunger, disease, and suffering is a real possibility for humanity as a whole, not just the fortunate few kings and priests. A truck with the power of 100 horses commonly does the bidding of people in modern society. In the past only very few could afford the luxury of the concentrated work of 100 horses. An opulent lifestyle has eluded most of humanity for much of history and still is out of reach for many on our planet.

Even people in developed countries, who feel secure that their way of life will continue in perpetuity, may not actually be as secure as they think. We inhabit a global society that is interconnected and heavily dependent upon other people and the environment to continue. Economic failure in one part of the world can ripple to countries far away. Our actions have global effects on the environment. We are at a unique point in history where the actions of humans can influence others around the world, because our food supply is international, we have weapons that can destroy the Earth, our effect on the environment has become global, and diseases can spread around the world.

Large-scale problems in the world are not necessarily a summation of local problems. To the victim of an auto accident, a rape, a robbery, or a beating, that is the worst problem. A starving person is unlikely to worry much about nuclear warfare. When someone close to us dies, it is the worst problem. At the time you or your loved ones are suffering such things, it is impossible to consider abstract issues of what could be the worst problems for others.

People everywhere are clamoring for your attention and trying to convince you that their issue is the most important. They may be sincere, they could be self-serving, and they are probably sincerely self-serving. The emotional appeals used to capture our attention are deeply personal. We see images of suffering people, and people try to convince us that the world will collapse if our beliefs do not change. Some threaten us with fire and brimstone, or at least hell on Earth, if we do not heed the cry of those who proclaim doom. In the face of sensationalism, we need a method to determine which are the worst problems based on empirical evidence, not emotional appeal. Only after we have identified causes of these problems do we have any hope of finding solutions.

There can be a multitude of causal levels for any problem. For simplicity, we can place causal explanations for the biggest problems on a continuum between two extreme levels: the immediate (proximate) cause and the root or underlying (ultimate) cause. A rational strategy is to first identify the proximate causes for the problem and then trace back to the ultimate causes. Then, we can analyze the chain of causality and formulate solutions for actions at the most efficient points. For example, it is generally better to treat the ultimate cause of a disease than the proximate; taking aspirin for a fever is less effective in the end than taking an antibiotic to kill a bacterial infection that is causing the fever. Ultimately taking steps to avoid bacterial infection in the first place is the best defense against the disease.

The ultimate cause for many of humanity's big problems is human nature. Realistically, basic human behaviors such as the urge to reproduce or the desire to attain comfort are very difficult to change, so the solutions to the world's worst problems lie somewhere on the continuum between ultimate and proximate causes. Let me clarify with an example. Global nuclear war has the potential to destroy humanity and most other species on Earth. The proximate cause is nuclear weapons. The ultimate cause is that a good portion of humanity has a propensity for conflict and war as well as the possibility to make mistakes. In this particular case, only allowing enough weapons to exist such that Earth cannot be destroyed is more likely to be successful than trying to stop all wars and assuming all people dealing with nuclear weapons are perfect. We can imagine arms control and are slowly achieving it, but it is unlikely that all international conflict will ever cease. Alternatively, nutrient pollution is causing nuisance (even toxic) conditions in waters around the world. Lowering the rate of pollution (the ultimate cause) makes more sense than treating every water body that the pollution runs into (a proximate solution).

I do not deny the existence and need for attention to problems that have strong negative effects on just a few people. It makes sense to attempt to find a cure for a disease that only kills a few hundred people a year, as long as the effort does not cause the suffering of disproportionately more people or cost a disproportionate amount of resources that we divert from more pressing issues. This is not a hypothetical scenario. Developed countries spend more money researching diseases common in these countries than those in other countries, even though other diseases, such as malaria, kill far more people.

What I attempt here is to adopt a broader vision that transcends a single area of the world or the experience of one or a few people. My goal is to establish a general framework to explore humanity's most prevalent problems or those that threaten the continued existence of humanity to the greatest degree. I use science and facts to approach the issues. This allows me to address root causes and solutions in a logical manner after the relative degree of threat posed by each of the worst of the problems and their causes are established.

The ultimate goal of this book is to create a way to rank the leading problems on Earth and then discuss how to solve them or mitigate their effects. First, I discuss the moral basis for what I will consider a problem and some likely candidates. Next, I discuss the candidates in more detail. Then I describe an index that can be used to rank the problems and the actual ranks and uncertainties in the ranks. The next section synopsizes what other writers and people I have talked to think are the worst problems. The last chapters consider how we might solve the problems, what the solutions would cost, and the social, economic, and political aspects that need to be considered to find solutions. Hopefully, you will find this all as interesting as I have, and you will be inspired to work toward solutions.

Chapter 2
Global Problems?

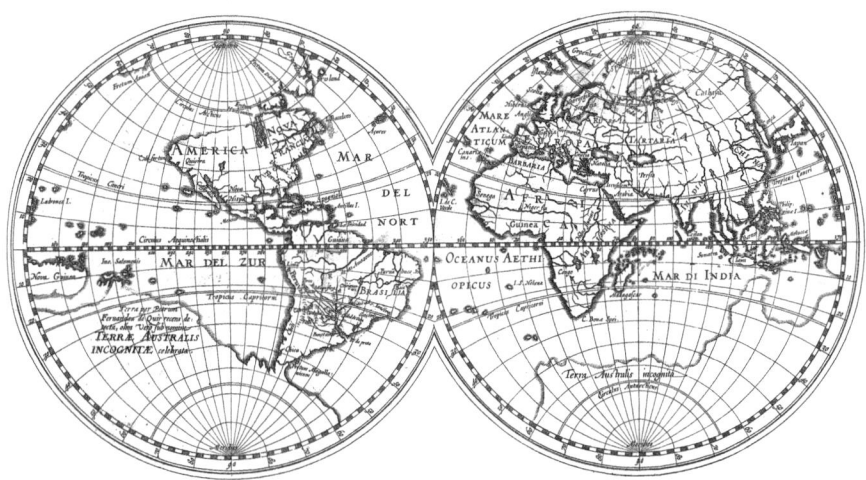

World map of 1612 including the discovery of La Austrialia del Espiritu Santo by Pedro Fernandes de Queirós

It is *possible* the worst will happen. An asteroid could slam into the Earth. Nuclear war could decimate the planet. Disease could sweep the globe. Social infrastructure could collapse. Do we know if our Earth that supports us will survive into the future? How likely are any of these things to happen? Should we worry about them? How do we rank the worst problems? The natural tendency is to worry about the worst threat to your personal existence or your children's happiness and survival. However, from an objective point of view, what causes the most human death and suffering now, and what will happen in the future? This is the central point of this book.

Many people only ask these questions a few times in a lifetime (usually as a young adult searching for meaning in life). It is not comfortable to think about these scary and depressing questions. The answers are not what we want to hear, and solutions can seem elusive. However, avoiding questions like this means never finding answers that could help people everywhere and alleviate actual and potential suffering. Here, I explore the questions in depth, try to provide pertinent information so individuals can make up their own minds about the severity of world problems, and try to find the common threads toward solutions.

© Springer Nature Switzerland AG 2019
W. Dodds, *The World's Worst Problems*,
https://doi.org/10.1007/978-3-030-30410-2_2

I take a scientific approach, but science says nothing about the importance of values. Therefore, the book also includes specific discussion on the moral underpinnings required to establish what a bad problem is, and then I attempt a non-biased approach to assess problem severity.

Taking the Bias Out of Confirmation

Some take an approach to understanding and reacting to conditions in the world by only considering facts that fit their worldview comfortably. That approach is destined to depart from reality. Worse, some make up "facts" to prove their point. This is the approach of some alarmist talk-show hosts, politicians, and "news" commentators. Fact checking be damned, they make their points, purporting to be objective and claiming anything that disagrees with them is "fake news." Incorrect news gets more interest on electronic media than true stories [1], but "fake news" is difficult to detect and identify [2]. Those who want their worldview confirmed cling to the untruths as if they are facts, even in the face of solid evidence that there is no truth to an argument [3]. Some people are more susceptible to fake news, and people who exploit this susceptibility and willingness to transmit the false information take advantage of this propensity [4]. This approach to understanding the world is a well-confirmed dimension of human behavior and is common.

The tendency of humans to filter facts to suit their preconceived notions is known as confirmation bias [5]. Many teenagers are quite good at filtering reality. Some think they are indestructible and will drive cars or motorcycles as if their personal invincibility was fact (and some just don't care). They will continue to drive this way even if they have friends that are hurt or die in accidents driving exactly the same way. They assume they have better reflexes or are more aware, even though the evidence is clear that driving too fast and recklessly leads to accidents.

The foundation of science rests on methods to avoid confirmation bias and preconception, thus combating the common problem of filtering facts and only accepting those that are comfortable or personally beneficial. The success of science is verification to the effectiveness of the approach of independent validation of facts and only accepting information or observations as true that agree with the bulk of other observations. Even though the Catholic Church attempted to silence Galileo, Copernicus was correct, and the Earth did and still does revolve around the Sun. In that example, the survival of humanity was not at stake, but faith in humans at the center of the universe was. Sometimes people don't want to believe facts because that is scarier than the possibility that their own worldview is not correct.

The successful scientific approach requires a change of worldview if the facts do not fit it. There are still large areas of human experience where there are no facts. In the realm where we do have scientific knowledge, the adaptive approach allowing reality to guide worldview has far more promise in providing effective solutions to problems than selective use of information. I use the scientific method here. The scientific approach is pragmatic and will most likely lead to accurate prediction of the consequences of our behaviors for most people. For example, I discuss how global control of chlorofluorocarbons stopped catastrophic global increases in harmful ultraviolet rays in Chap. 6. Numerous other examples also exist, such as vaccination to prevent disease and improved crop yields to prevent hunger. In this book, I will discuss the ways that social science has improved our ability to predict human behavior. The enlightenment provided by the scientific approach has led to most of the advances that have improved human standard of living and comfort. This enlightenment has led to greater happiness for the more fortunate in the world.

I construct the arguments in this book based on all the available facts related to the problems, arranged in their most logical order. When, in the future, new knowledge emerges that does not agree with some of the specific arguments presented here, my view of what are the worst problems, what causes the problems, and what are potential solutions will change. This is the rationalist approach of the scientific method and has more chance of long-term success than an ideologically driven approach.

I separate the ethical dimension of why a problem is "bad" and how to deal with the problem from the discussion of problem severity and causes. I present the specific moral principles used as the underlying basic assumptions of what might be the "worst" problems and what are acceptable solutions to those problems. Logically, all human suffering would end if all people ceased to exist, but morally, this is not an acceptable solution to the world's problems for me or most other people.

A scientific approach provides no way to prove that death and suffering are good or bad. Here I apply a moral stance that views suffering and death of people around the world as wrong and admits that the nonhuman natural world has intrinsic right to exist. My approach to the worst problems is to identify the problems experienced by humanity as a whole.

I started with the idea that most people do not want to think about difficult questions such as those approached here. Even thinking about such questions leads to charges of being a bearer of bad news, a proponent of "doom and gloom." A common response is to "shoot the messenger." Chicken Little said the sky was falling and it never did. The boy should not have cried wolf so many times. If the first fable is correct, then the messenger maybe should not be shot, but certainly should be ignored. If the second is correct, then the wolf does come eventually. Will the wolf really come or maybe is here already? I try not to raise false alarm, but do not want to stick my head in the sand. I assess the worst problems by listing the specific scientific facts involved and predictable outcomes of our actions, because the facts are the facts. Again, facts form the basis of wise action and paths to solutions.

What are the potential threats to humanity, and can all on Earth enjoy a secure and healthy existence? Does this secure and healthy existence require a standard of living for everyone that is comparable to that of much of the developed world? These questions form the core of my book, and I will approach them generally and then individually.

When I wrote my book on global environmental issues, *Humanity's Footprint* [6], it started with a list of what I thought, at the time, were the world's worst problems. I based this list on my general feeling about the biggest threats, but I did not use rational criteria for determining the worst problems. I will try to create a more logical list now. The list is mostly human-centered (anthropocentric), but does leave open the possibility of other gauges of severity of problems (e.g., the rights of species other than humans to exist). First, I discuss more thoroughly the intrinsic moral assumptions behind identifying the worst problems, because these assumptions form the basis of ranking various problems.

Moral Assumptions

Addressing the world's problems requires placing some moral weight on defining what actually constitutes a problem. We can use this moral structure as a common foundation for the arguments about what are the worst problems and how to rank them. Clear and neutral assumptions make moral questions accessible to the scientific approaches of analysis of facts and quantification [7]. My basic assumption is:

> Death or suffering of any person, now or in the future, is weighted equally across the world. The specific nationality, race, class, gender, sexual orientation, or religion of a person does not mean that they deserve death or suffering any more or less than any other.

No single person deserves to suffer more because of where they live or what they believe (unless what they believe leads them to behave in ways that harm other people). This is the basis of humanistic morality [8].

This moral structure goes against some aspects of fundamental human nature; individuals do not naturally view all human suffering as the same across the world. Awarding intrinsic rights to species other than us is even less basic to human nature. Suffering and death of people closely related to us influence us much more than when unrelated people are harmed a continent away. We perceive our own pain as greater than that of others, particularly if those others are of a different race [9]. Different parts of the human brain fire when observing another person in pain than those that fire when we ourselves are in pain [10]. This propensity of humans to protect themselves and those close to them has a strong evolutionary basis; it has been essential to survival and success of the human species for most of our evolutionary history.

If a sniper kills a dozen people in your own town, it makes a much bigger impression than if an earthquake kills 100,000 in another country on the other side of Earth. This makes sense because dangers close to you are more likely to hurt you or your relatives than those far away. Caring more for those close to you is logical as a hardwired trait or a culturally learned behavior; it increases the survival and growth of families when they stick together against the rest of the world. Still, having a global perspective that does not weight the suffering of any individual more heavily than another avoids the unfair and unequal weighting that occurs from assuming one group of people is more deserving.

> Equality means everybody has the same right to live a full life and in a healthy environment, and nobody should suffer disproportionately because of factors outside of their own control.

With this moral basis as a common starting point, I build a structure to decide what makes one problem worse than others. One issue that immediately comes up related to severity of problems is how we equate suffering to death. Again, this is a moral issue with no scientific answer. There are several ways to view this. Let us say you could know with absolute certainty that you could choose to live 30 years and suffer in terrible pain the whole time or die now. What would you do?

Another way to look at it is from the view of the leader of a group of people. Assume there is another group of people that is attacking you to capture and send your people into slavery (keep most people alive, but suffering). Also, assume you know how many casualties among your own people there will be to defend yourselves. How many people should you send to fight and potentially die to protect more people from slavery?

People in mental anguish and physical pain make the decision to end their suffering and choose to die by suicide rather than live. How does a terrorist bomber view their action? The suicide bomber perhaps thinks that the loss of their life is worth making the world a better place for other people. This balancing of suffering and death happens all the time, but putting a general absolute value to it is a moral decision. I will revisit how comparing the harm from death to suffering alters what we view as the worst problems.

> Furthermore, I assume that the natural world, apart from humans, has an intrinsic right to exist.

Human suffering and death is not necessarily the only aspect that we can consider symptomatic of the world's worst problems. I am deeply worried about the fact that half the species on Earth will go extinct in the next generation. There are practical reasons to be concerned about this extinction because we might cause the loss of a species that is central to system stability. Loss of the wrong species could cause the collapse of the ecosystems that support us. In addition, humanity could wipe out species that have the genetics that code for the next cure for a common cancer, improved productivity of food crops, or any of many other uses to which people put plants and other animals.

The idea that nature has an intrinsic right to exist is an approach taken by many traditional conservationists, as prior to the conservation movement most people only viewed nature as something to be exploited [11]. While extinctions of other species could have negative consequences for humanity, what truly bothers me about extinctions is fundamentally a moral issue. I do not believe that humans have the right to wipe out other species for the benefit of individual people. I understand that some species go extinct naturally, but am unwilling to accept that the human "need" for shopping malls, roads, and material gain above that required for a reasonable, albeit moderate, standard of living outweighs the rights of a nonhuman species to survive. The utilitarian view runs counter to my moral view. Why does it matter if a species goes extinct? Do lower animals even have feelings? Plants certainly do not. Would it be a disaster if the mosquito species that carries many human diseases (*Aedes aegypti*) were to go extinct?

Not everyone will agree with my approach here. Some people believe with absolute conviction that the worst problem in the world is that people do not believe in their version of God. Some believe that specific behaviors are so terrible and blasphemous that they are the worst problems in the world. Some people believe that the world would be a better place with no humans at all (and advocate for no reproduction). All of these positions on problems are subjective, and we cannot use facts and logic to assess their severity.

Moral issues are impossible to settle with rational argument. Therefore, I simply state my moral assumptions, but do not spend much time arguing for them. For example, one religious person could believe that God made the Earth for humans to use how they see fit. Another believes that it is desecrating the creation of God to cause the extinction of other species from arrogant, self-serving human activities. Either one can argue it is their belief, but there is no way to prove scientifically which view is correct. As I lay out the arguments and the quantitative basis behind those arguments, I hope that, as a reader, you can adjust my proposed approach to match your own moral basis.

It is not possible to design an experiment to test morality, religious belief, what is art, or many other realms of human culture. Therefore, I am starting the arguments here by specifically defining my yardstick for determining what makes one problem worse than another. By picking the death and suffering of people as my yardstick for severity of problems, I am picking not only what I believe is important but also what I believe that many just and moral people around the world think is very important as well. By assuming that all people are of equal importance, I am making the same assumption that arose from the Enlightenment and inspired the preamble to the US Declaration of Independence. "We hold these truths to be self-evident, that all men are created equal." Of course, I take the broader view by assuming "men" means "people." By giving other organisms on Earth an intrinsic right to exist, I am stepping further away from an anthropocentric view, but this is something I feel is important. The readers can decide for themselves if the moral basis I have chosen works for them, and I will try to spell out which consideration applies where, so the logical structure can still be useful even if you have a different moral framework you want to apply.

Given the basic moral issues discussed above, I start the discussion on the world's worst problems with my list of potential reasons to consider a problem one of the worst: (1) causes of death and suffering now, (2) potentials for catastrophic death and suffering in the future, and (3) threats to the long-term stability of society and the environment that supports us. Then I will try to synthesize these into a final list of problems to consider. I will cover each of these points in the next sections.

How Are People Dying Unnecessarily Right Now?

The first gauge is the number of deaths that actually are occurring. The World Health Organization lists the top 10 causes of deaths on Earth as of 2016, in millions of people who die each year. Causes are ischemic heart disease and stroke (15.2), chronic obstructive pulmonary disease (3.0), lower respiratory infections (3), Alzheimer's and other dementias (2.2), lung cancers and diabetes (1.6), road injuries (1.5), diarrheal diseases (1.4), and tuberculosis (1.3). We have clearly made some progress on some problems such as HIV/AIDS and low birth weight, as these have fallen off the top 10 list in recent years. Are all these deaths the same? Are some inevitable consequences of old age?

A list of causes of death does not provide a simple gauge of worst problems, however. A number of these deaths relate to diet and lifestyle choices. Specifically, for all people on Earth (again according to the World Health Organization), the top 10 risk factors, in descending order of severity, are high blood pressure, tobacco use, high blood glucose, physical inactivity, overweight and obesity, high cholesterol, unsafe sex, alcohol use, infant underweight, and indoor smoke from solid fuels. While many of these can have a genetic component, many have a component of risk derived from behavioral choices. Particular exceptions to the idea that people can choose the risks that lead to mortality are childhood underweight (undernourishment) and exposure to indoor smoke (inadequate ventilation and fuel for heating and cooking, as well as secondhand smoke).

If an individual chooses to smoke and knows that it is dangerous, can we consider their death (other than health costs and suffering of loved ones) as important as the death of a child who does not have access to adequate food? About 80% of the health-related deaths are from cardiovascular disease, the number one killer on Earth, and many of these could be prevented by healthy diet, physical activity, and avoiding smoking tobacco. As the leading cause of death, it is an extremely important problem, but can society be responsible for controlling people who do harm to themselves? If all people must die of something, is the most common cause of death of older people a worst problem? Several of the top 10 risk factors are particularly likely to cause death of older people. There are several ways around these particular issues in order to approach what we might consider the worst sources of mortality in the world.

One way to account for deaths that have greater effects on older people is to consider age-adjusted death rates. This adjustment accounts for the number of years of life that is lost over the normal life expectancy. Lawsuits take the adjustment approach to calculate compensation rates for a death or injury. Adjustment accounts for life expectancy and future earning capability. Age adjustment of death rates allows some people to argue that heat-related deaths in large cities related to global warming are not that bad and that these deaths mostly involve older people who do not have that many productive years ahead of them anyway. Part of me believes the approach mimics the alleged Inuit practice of leaving a grandparent on an ice floe in hard times (a practice that is not exactly factual [12] and may not even be possible in the future with global warming). In any case, my belief that the age adjustment approach is clinical, if not outright cynical, leads me to be cautious in my logic about how death rates relate to the world's worst problems.

Personal experiences are relevant to why we might not want to consider a simple tally of deaths as a final indicator of severity of global problems. Certainly, you can provide your own similar experiences. My 93-year-old great aunt died in her home according to her wishes after a good life. It is difficult for me to worry about what the official cause of death was. I miss her, but she lived longer than most, and death is inevitable. She was to the point that she was ready to be done with it all. Still, her death will be part of the statistics of morbidity. On the other hand, the son of my friends died by suicide. The suicide of this young man had a stronger emotional influence on me because this wonderful young man possibly could have had a productive and happy life with the correct intervention. Do the two deaths count the same when tallying the global impact of death? Local cases color our impression of large-scale trends. So, I will continue in the vein of large-scale statistics by assuming that it is more useful to consider deaths that are preventable by society as a whole and are not controllable by the people who are actually dying.

Given the desire to find a way to tally deaths by preventability and inability of the victims to avoid death, the best way to consider death rates could be to count only child mortality. In this case, lifestyle choices of the child are causative of very few of the deaths. Around 10 million children under the age of five die each year on this Earth. Most of these deaths occur during or soon after childbirth, often related to inadequate prenatal care or health conditions during childbirth. Undernutrition is the underlying cause of about 1/3 of these deaths. Having too little to eat is certainly not a lifestyle choice for most people. A substantial number of these deaths are due to preventable or treatable diseases such as diarrhea. I argue that these numbers indicate that undernutrition and inadequate sanitation leading to unsafe food and water are major problems on Earth because of the number of innocent people influenced. Disease and hunger rank high on the list of involuntary and mostly preventable causes of death.

About 15 million people die each year, but there are about 40 million induced abortions each year. I do not believe that abortion is morally equivalent to death, but many disagree with my belief. Drawing lines is difficult. Does birth control also count? Is it a problem if people do not try to have all the children that they can? According to the World Health Organization, abortion rates are falling globally. Conditions that lead to better maternal health and lower rates of child mortality (e.g., higher standard of living, prenatal nutrition, education for women, and access to contraception) also lead to lower rates of abortion and of mortality from unsafe abortions. Thus, solving childhood hunger and disease should also decrease the number of abortions. We are also making progress on that front, but much is left to be done [13].

What Might Lead to Future Global Catastrophes?

Any self-respecting alarmist could list numerous potential causes of widespread death. Disaster and James Bond movies seem to be particularly adept at making up worst case scenarios for the majority of the Earth's population, but in reality, the potential causes are less numerous. While recent natural disasters have been devastating to some regions, they are less than globally catastrophic for the purposes of this discussion.

A number of people might argue that war is one of the world's worst problems. World Health Organization data suggest that in 2002, wars caused about 0.3% of the deaths globally. War does not make the top 10 list in most years. However, World War II caused the deaths of about 2.5% of the Earth's current population over a period of about 6 years. To put this in perspective, that was about twice the current annual total death rate. Half of the deaths were civilians; about 1/5 related to disease and famine. However, this was, hopefully, a global anomaly, and viewed over the long haul, war is not a top cause of mortality on the planet. Stated differently, if we spread that huge mortality from World War II across the last 65 years, it contributes about 1/30 of the death rate; thus averaged over time, war is not in the top 10 causes of death currently.

We have known since the beginning of the Cold War that global nuclear exchange could wipe out all people and most plants and animals on Earth. The insanity of having enough nuclear weapons to erase most life on Earth not just once, but many times, did not stop the United States and the USSR from amassing ever larger stockpiles of nuclear weapons over the decades. Many of these weapons still exist, and additional countries are adding to their nuclear arsenals. Now we know that even a relatively limited nuclear exchange could be catastrophic. I have no doubt that the continued potential for nuclear war is a strong candidate for the list of the world's worst problems. Maybe that is because I grew up during the cold war, a time of bomb shelters and duck-and-cover drills. For now, nuclear war remains on the list.

Another potential source of future catastrophic death is the emergence of a fatal incurable disease, probably viral, that kills a large portion of the Earth's human population. Several waves of disease have occurred over the last decades. The recent H1N1 pandemic, Zika, Ebola, the avian flu, and West Nile virus show that new diseases are frequently emerging and that they can spread rapidly around the world. Had these diseases been a bit deadlier and/or more contagious, the outcome could have been extreme suffering and even death for a large portion of the world's population. Given the possibility of bioterrorism and designer diseases, the threat is greater than ever before. Because a serious pandemic of a deadly disease has so much potential for global destruction, this is also on the list as potentially one of the worst problems.

Humans could destroy the protective layer of ozone high in the Earth's atmosphere. This layer keeps most of the dangerous ultraviolet rays from reaching the surface of the Earth. If this layer did not exist, all plants would be sunburned, animals could not survive for any length of time, and humanity would starve. The use of chlorofluorocarbons as propellants and refrigerants started to destroy the ozone layer, and the international community agreed to limit their release. However, recent scientific developments suggest that the ozone layer is susceptible to other causes of destruction. Specifically, increased use of nitrogen fertilizers that will be necessary to feed the growing human population stimulates microbiological activities that create ozone-destroying chemicals. Thus, the potential for catastrophic global death and suffering is high, so I will add this to the list along with other aspects of the global environment as part of the world's worst problems.

Many people consider global warming and associated climate change to be the greatest global problem. While the mortality and suffering due to this warming are a bit harder to predict, the problem deserves further consideration, and I will discuss it under the general topic of global environmental issues, including ozone and species loss.

Some potential additional globally catastrophic events include collision with a large asteroid, a super volcano, gamma ray bursts, and total breakdown of societal order. For various reasons, I will discuss them briefly in the next chapter.

How Are People Suffering Now?

The discussions on death causes also give clues as to who is suffering now. Some of the top 10 causes of death also entail a substantial amount of suffering and indicate that numerous other people on Earth, while not dying from these causes, are experiencing misery from them.

Communicable diseases lead to a large amount of suffering. HIV/AIDS, malaria, and tuberculosis cause a substantial portion of the deaths in the world. Many of these deaths are occurring in people who lack adequate medical treatment for the disease, suggesting that their degree of suffering associated with these diseases is substantial. Given that many of these diseases are preventable or that it is possible to alleviate the suffering from these diseases, I consider disease as one of the worst problems because of the suffering and death it causes.

According to the United Nations World Food Programme, almost 1 in 6 people on Earth do not get enough food to be healthy and lead an active life. While starvation is not one of the top 10 direct causes of death in the world, it is a contributing factor in many deaths. Of course, the fact that 1 in 6 people on Earth, over a billion people, is hungry most of the time is clearly a widespread cause of suffering. These people are particularly susceptible to economic downturns and political unrest. Factors influencing food prices, regional wars, and drought can all cut off supplies.

One quarter of the malnourished people on Earth are children. The lack of food is clearly not their fault. Given considerations of suffering and the fact that hunger can be controlled, I think that world hunger is one of the world's worst problems. The fact that so many children starve while humanity is producing enough food to feed them now is truly one of the great tragedies of our times.

Risk Analysis and the Future

It is difficult to know how to weigh future problems that might occur against those that are happening now. Preventable diseases and hunger cause over a billion people on Earth to suffer every day. Global nuclear war would likely eradicate humans from the planet and lead to tremendous amounts of suffering in the process. How do we play these two scenarios off against each other? I will take up my approach to analyzing severity of problems in detail further on in the book, but I raise the issue here so readers can be thinking about it as they go through the individual problems.

People make calculations that weigh personal suffering and chance of death which influence their behavior every day. Death is the ultimate penalty, according to the majority of people. We are like all other animals and fight to survive if we are threatened. Some diseases cause us suffering but generally not death. Thus, we are willing to be exposed to someone with a common cold, but highly reluctant to risk exposure to the Ebola virus.

People are notoriously bad at understanding the true risks of many activities. Those in the United States who have been alive long enough to remember the results of the attacks on September 11, 2001, know the effect this had on all our lives; the attacks changed society in the United States. Terrorists killed approximately 3000 people. This level of fatality is about equal to the number of people who die in automobile accidents in the United States in 26 days. An immediate concern makes perfect sense, as this could have been the first strike in a larger scale attack. However, as time went on, it became clear that the danger in the United States was not likely to cause more deaths than automobile accidents or maybe even cases of domestic terrorism. For example, in 2015 in the United States, there were 44 terrorism deaths, 28 war deaths, 15,696 homicides, and 35,398 automobile accident deaths. Globally there were 38,422 terrorism deaths, 97,496 war deaths, 437,000 homicides, and 1.25 million automobile accident deaths in 2015 [8]. Still, in the United States, the public takes little time to worry about automobile accidents, even though legislating safer automobiles and tighter regulations, particularly lower speed limits, could save many lives. In contrast, much press coverage and worry continues over terrorist attacks, even though the number of deaths is low relative to some other easily controlled causes. Perhaps terrorism is a sign of future societal collapse, but predicting the likelihood of this is difficult, and I have not seen any method for accomplishing such prediction. Thus, depending on circumstances, humans do not always calculate risk in a rational fashion.

The arguments in the following chapters will attempt to gauge risk based on facts, not fears based on feelings. It is true that sometimes a "gut feeling" will predict the future better than an expert opinion, but not often. Later, I will discuss how a group of experts can actually make better predictions than individual experts can (they have a sort of group intuition that exceeds individual predictive abilities). For now, we will stick with quantifiable risks based on facts.

The Emotional Bits

Being the bearer of bad news is not a popular stance. Most people would rather avoid difficult questions and issues. When I talk about the big problems to people, many of them shut down or try to change the subject. Some joke, and I am guiltier of this than most.

I have been involved with research on the frog extinctions in Panama. These extinctions are resulting from a fungal disease that is sweeping through the high cool mountains of Central America, causing the extinction of tens to hundreds of species there. When I start talking about this project, people often go quiet and then quickly change the subject. Most people would rather talk about the weather, shopping, or a local sports team than consider the thorniest issues in life.

Our unwillingness to face horrible possibilities in reality is somewhat baffling to me. I know violence, fear, and danger fascinate many people and they watch it for entertainment. People are willing to pay to go see disaster or horror movies, yet are not willing to contemplate the equivalent scenario in reality. People thrill to watch fictional assassinations by organized crime figures, but do not want to consider the deaths that occur every day.

Sometimes it can be dangerous to take the world too seriously. A friend decided she should not get old. She thought that her ecological footprint (the effect that each person has on the environment) was too great and that the money she had would be better spent on her favorite causes than on taking care of herself in old age. She retired from her job and quit volunteering. The police received a call at 4:00 a.m. on a cold November morning that there was a body in the local cemetery. When they arrived, they found her, shot in the chest. She called the police because she did not want the students of the nearby high school to find her in the morning. Before, she had sent emails to her friends with details on how to deal with her possessions (all boxed and labeled in her house) and why she had done what she had done. She was an environmental and social activist and had been dealing with depression prior to her death. Who can really understand the actions of another? It appears that, at least in part, dealing with the world's problems simply became too much. This makes me think that even considering the worst problems is maybe not such a great idea.

Still, I would not write this if I did not hope in some small way that it might make a difference. The margin between disaster and sustainable survival may be narrower than we think, and I am optimistic enough to think that we might be able to forestall disaster with appropriate scientific and cultural understanding and action. Maybe somebody who comes up with a solution to a pivotal world problem is reading this and will be inspired toward finding a solution.

I am going to try to keep emotion and fact separate in this writing. This means that when something is a fact or logical, I will say "I think that…," but if something seems morally or ethically correct to me, I will say "I believe that…." A strong belief may seem to have more weight than a logical thought, but facts are more difficult to dispute than beliefs, and belief is subjective. You may or may not believe what I say here, but I think that facts are facts.

The problems are so extreme that sometimes the only options seem to be to laugh or to cry. I cry often. To keep going, a joke is what I need sometimes. Even Holocaust survivors report that humor was important for their survival. Please excuse my pathetic sense of humor, and I hope you get the irony. My friends maintain that when people laugh at my attempts at humor, it only encourages more lame jokes.

The Problems Considered in This Book

My core list of worst problems is based on the arguments explained above. The main suspects are, in no particular order for now, (1) hunger, (2) disease, (3) nuclear weapons, and (4) global environmental impacts. I have picked these four general categories because they are not controllable by individuals and they currently, or have the potential to, cause death or suffering of the greatest number of people around the world. Still, other problems could have catastrophic consequences. I will consider each of these problems and then will discuss each of these problems from the above list in detail with respect to their effects on humanity and the proximate and ultimate causes. In addition, I will consider some more catastrophic scenarios in the next chapter. Following this, I will categorize how each of these problems and subcategories of problems ranks with respect to current and future death, suffering, and ability of Earth to support humanity. This categorization proceeds sections on potential solutions to the identified problems.

Chapter 3
Apocalypse

The Apocalypse, Albrecht Dürer. 1496–1498

W. Dodds, *The World's Worst Problems*,
https://doi.org/10.1007/978-3-030-30410-2_3

This chapter is the stuff of disaster movies. The apocalypse could be lurking around the corner waiting to pounce. Chicken Little will be right; the sky will fall. Events will vindicate the boy who cried "wolf." The problems discussed here are either very unlikely to occur, difficult to substantiate if they will occur, impossible to do anything about, or some combination of the above. Storing large amounts of food to offset the potential crop failure is about the only insurance against most of these threats. Guaranteeing food security would require global cooperation if the world was to be fed. These are all potential catastrophes; if we are going to consider all possibilities, at least we get the thrill of considering all of the worst at its most spectacular. In Chap. 8, I will assign actual probabilities to some of these more likely events and compare that to the leading candidates proposed in the last chapter.

Asteroid

Asteroids or comets can cause catastrophic destruction of much life on Earth, as proven by the impact that caused the extinction of most of the dinosaurs about 65 million years ago as evidenced by the geology of the Cretaceous-Tertiary boundary. Dinosaurs disappeared from the fossil record at the same time that a global layer of iridium (an element most likely originating from a meteorite explosion) occurs in that same fossil record. Scientific data, including the remains of a huge impact crater, support the idea that a 6-mile-wide asteroid crashed in shallow water over what is now the Yucatan peninsula.

The explosion created an immediate massive tsunami (potentially with waves hundreds of feet high washing inland in coastal areas around the Gulf of Mexico). The impact injected aerosols of sulfate from the ocean water and underlying limestone high into the atmosphere, potentially leading to acidic rain for more than a decade after the explosion. Most destructively, fine debris ejected into the atmosphere and almost completely blocked the Sun for around a year, plunging the Earth into global winter. This cold period probably led to the death of many species of plants on land and microscopic plants in the ocean, as well as most of the animals that relied upon those plants. Not only did many plants freeze, but also the low light inhibited the photosynthesis the few remaining plants required for survival. Scientists also hypothesize that larger fragments from the explosion fell around the Earth causing numerous firestorms. High concentrations of oxygen in the atmosphere during the Tertiary period would have helped fuel these fires. The fires, coupled with rotting vegetation after a dark year, probably released enough carbon dioxide and other gasses to cause a decades-long global warming (greenhouse effect) once the global winter abated.

A similar collision with an object from space today would cause catastrophic effects for most people on Earth, either causing complete extinction of the human race or leaving few scattered survivors living under harsh primitive conditions. Such an event could occur now, during the newest geological age of the Anthropocene, when people rather than dinosaurs dominate the planet. The largest collision with an asteroid or comet in recorded history was the Tunguska event in an uninhabited area of Russia in 1908. Just after a sunrise in June, residents north of Lake Baikal, in Eastern Russia, saw a streak of intense blue light moving across the sky that was too bright to look at directly. A huge explosion in the distance followed this streak. The flash of light caused a short and intense heat burst. Then, a shock wave hit that was so strong it knocked people over and broke windows a hundred miles away. The explosion was slightly smaller than the largest nuclear bomb ever exploded.

Those closest to the explosion reported tree tops snapping and burning, ignited by the blast of superheated air that followed the shock wave. Farther away, the searing wind damaged crops and other plants. The explosion was intense enough to cause an earthquake with a Richter magnitude of 5.0. Curiously, there are no written accounts of exploration of the site immediately after the explosion.

A Russian team, led by the scientist Leonid Kulik, in 1927, undertook the first official exploration of the blast area. This group found an area of 40 by 30 miles with uprooted trees flattened with roots toward the center. The team was surprised there was no crater and they found no meteor fragments. They surmised the object had exploded before it had hit the ground. By the 1960s, Russian scientists calculated it had exploded with a force of approximately 1000 times the destructive force of the Hiroshima bomb (about as strong as the fifth most powerful atomic weapon ever tested).

A large meteor exploded above Chelyabinsk, Russia, in February 2013. This object exploded with the force of 20 Hiroshima bombs. The shockwave damaged buildings and shattered glass, injuring over 1000 people. The object was over 10,000 tons and 55 feet wide and exploded about 20 miles above the ground. Astronomers estimate that such objects collide with Earth about every 100 years. Interestingly, a larger asteroid (estimated at 100 yards across, 143,000 tons) passed very close to Earth on the same day. Had this asteroid collided with Earth, it could have devastated an area of 750 square miles.

So how likely is a collision of an astral body with Earth to cause human death and suffering? Where they would hit is random, so most would land in the ocean. The smaller the object, the larger the probability one will strike the Earth. The largest asteroids that could cause planetary extinctions on Earth may strike every 100 million years. Smaller objects capable of explosions the size of the bomb that leveled Hiroshima may strike the Earth every few years. There are no records of a human fatality from a meteorite in the last 1000 years. Ancient Chinese records suggest such deaths have occurred in recorded human history.

An object as large as that causing the Tunguska event could kill millions of people if it hit in the wrong place. Such objects are "city killers." Astronomers have calculated that such an impact will occur on average once every 2000–3000 years. However, most such objects would fall in the oceans that cover over 2/3 of the Earth's surface. Cities of more than 75,000 people cover 3% of land, so urban areas cover only about 1% of the Earth's total surface; only 1 in 100 objects large enough to be "city killers" striking Earth, on average, would destroy a city. Thus, the probability of an asteroid destroying a city would be roughly once every 0.2–0.3 million years. These are good odds for most people on Earth.

Astronomers published improved estimates of the probability of asteroids colliding with Earth based on detailed surveys of space in 2004. An asteroid capable of causing global disruption by killing many people and destroying our ability to produce food would have a diameter greater than about a half mile. If such an object struck Earth, it would release the energy of about 10,000 atomic bombs of the size that destroyed Hiroshima. An object this large or larger is expected to collide with Earth roughly once every half billion (500,000,000) years. This is about 1000 times longer than the length of time between now and when our human ancestors diverged from chimpanzees. A regionally destructive collision with an asteroid with a diameter of 900 feet happens roughly every 50,000 years. Such a collision could have global impacts on climate due to the dust ejected high into the atmosphere, interfering with successful crop growth, and destabilizing human societies. Such an event could have happened 10 times since hominids started walking the Earth.

As of 2011, scientists estimated they had identified 90% of the near-Earth objects (asteroids) larger than 1 kilometer (0.6 miles), and none of them is likely to hit Earth over the next century. There are quite a few asteroids smaller than 1 kilometer that are still large enough to wipe out a city and have regional effects. We have identified an estimated 5% of these, and the goal is to identify 90% by 2020. Thus, the chance of Earth colliding with a globally destructive object over the next 100 years is slim.

There are large comets that we have not identified, and one could collide with Earth. More surveillance is necessary to find such an object before it collides with Earth. NASA has a goal to track all objects greater than 0.62 miles in diameter. There are many objects large enough to be "city killers" that could collide with Earth that we have not yet have cataloged (only 1% or less are known). The smaller the object is, the more difficult to detect. Astronomers did not identify the meteor that exploded over Chelyabinsk, Russia, prior to its encounter with Earth. Another asteroid passed close to the Earth in February 2013 that was 10 times larger than the meteor that exploded over Chelyabinsk. Astronomers only detected this object a few months before it passed near us. Had it been on a direct collision course with Earth, we probably would not have had time to do anything but evacuate an area where we predicted the asteroid would hit.

We would have to deflect a large object threatening Earth once we detect one on a path toward us. The movies, *Deep Impact* and *Armageddon*, both depicted a manned spaceship landing on an asteroid bound for Earth, and placing a nuclear bomb in the center of the unstable rocky object to blow it up. Unfortunately, such an approach with a real asteroid could just lead to many little asteroids that could cause as much or more damage than one big one. An even more frustrating scenario is that the gravity of the small objects would pull them back together before entering the Earth's atmosphere, and reform the original mass.

Potentially viable alternatives to avoid destruction by an asteroid include a gentle nuclear blast to nudge the object into a different trajectory. With enough time, we could paint one side of an asteroid with titanium. Titanium is white and reflects light. The dark side of the object would absorb the radiation from the Sun and reradiate it into space as heat. This would ever so gently push it to an alternative trajectory. Black paint on one side of a reflective object would have a similar effect. We can see this effect in the little black and white propellers vacuum-sealed in glass globes that spin in the sunlight (Crookes radiometers). The lack of an atmosphere drops the friction such that the pressure of the light particles can cause the propeller to spin. A third approach would be to slam an object into the asteroid to slightly deflect its path. Such an approach will be tested by NASA in 2022. Given the vast distances in space, a nudge is all it takes to make the difference between collision with Earth and a complete miss.

One country could accomplish the astrological cataloging of objects and subsequent response to protect Earth from a threatening object. However, protection that is more effective would involve international cooperation. International cooperation is the norm in astronomy. Many enthusiastic amateurs around the world have cataloged new objects. The largest space programs now involve international cooperation; thus, such avenues are already in place if a rapid response to a newly detected dangerous object is required.

Given the remote chance of such a globally devastating event occurring in our lifetimes, and a good chance we can deal with the problem because we will detect it in time, I will not consider the idea in much more depth. The concept is interesting though, because it does underscore how some events can cause global problems and it is the first time of several we will consider how disruptions of global climate can threaten human society.

Reversal of the Earth's Magnetic Field: My Cell Phone Won't Work!

The center of the Earth is a solid mass of iron, and molten iron churns around that core. The movement of the molten metal creates the Earth's magnetism. Given that the flow paths of the molten iron can change, the Earth's magnetic field can change as well. The Earth's magnetic field reverses occasionally, with times between reversals ranging from 5000 to 500 million years. The last flip was about 750,000 years ago.

The magnetic field of Earth shields the surface from the solar wind, a stream of charged particles produced by the Sun. This is potentially damaging radiation. It is possible the field would completely disappear during a flip, but computer models indicate we would still have a magnetic field, just more complex with many north and south poles in different parts of the Earth. The many magnetic fields could still shield Earth from the solar wind.

We do not understand all the biological effects of a switch of magnetic fields. Several switches have occurred since hominids evolved, indicating that biological effects on humans, at least, were not extreme enough to cause extinction. However, for most of human history, the dependence upon computers was not quite so great.

Past solar storms tell us something about what might happen during a magnetic reversal [14]. In 1859, a huge solar storm disrupted the Earth's magnetic field. The Aurora Borealis and Aurora Australis were visible near the equators. The storm knocked out telegraph lines, exploding batteries and sending sparks flying. A line from Boston to Portland Maine still worked even though it was not attached to any battery power; the storm provided the electrical power and sent electric currents through the line. This disturbance lasted 11 days. Today's electronics would be even more susceptible to the solar wind; loss of the magnetic field could damage computers, cell phones, and other electronics. The loss of global communication and computers could have disastrous effects on human economics and social stability. The effects could be worse in developed countries that are much more reliant on electronic communication and computers in day-to-day life.

Several types of animals use magnetic fields to guide them in their migration or other movements. Animals that rely upon magnetism include sea turtles, bats, migratory birds, cockroaches, spiny lobsters, salamanders, trout, whales, and even bacteria. All these species were present through several magnetic reversals and survived, indicating the shifts were not hard enough on most of them to cause their extinction. Many of these organisms use other senses to navigate and probably are able to re-calibrate their compasses with other cues.

It is also possible that reversals caused the protective layer of ozone in the upper atmosphere to thin and increased the amount of damaging ultraviolet radiation reaching the surface of the Earth; the data are inconclusive if past events had this effect and seriously harmed life.

Magnetic reversals are relatively rare, and it is probable the effects would be limited with respect to human health and survival. Given that we can do very little to prepare ourselves to deal with the negative effects of magnetic reversal and there is nothing we can do to prevent it from happening, such an occurrence will not be considered further in this book.

Super Volcanoes: Human Sacrifice Is Not a Solution

Every year or so, people must evacuate large areas or even entire islands along the Pacific Rim because of volcanic explosions. One can imagine if an entire island—your entire world—were threatened. Is it possible that a big enough volcano could have strong negative effects on the entire Earth? Mega-eruptions could well have caused the greatest extinction of all, the loss of 90% of all marine species on the planet about 250 million years ago at the Permian-Triassic boundary. Scientists are not certain of the exact mechanisms of extinction; it could have been huge amounts of sulfuric acid and acidic rain that fell around Earth after volcanoes spewed it into the atmosphere, or it could have been massive coal fires triggered by the explosions.

About 70,000 years ago, a volcano in what is now Indonesia exploded and caused a global winter for the next 5–10 years leading to massive deforestation in large areas [15]. This huge volcanic explosion left behind a crater that became a lake 100 feet deep and 62 miles long. This explosion possibly knocked down the global human population at that time tenfold, almost leading to extinction. Scientists think the explosion decimated the human population based on analyses of genes, although this interpretation remains controversial [16]. The bottleneck caused by an extreme decrease in population with just a few reproductive individuals leading to the entire surviving population leaves a distinct genetic trace. If such a massive explosion occurred now, the death toll and human suffering would be extensive.

Scientists classify a volcano that explodes and injects more than 1000 cubic kilometers (240 cubic miles) of Earth's crust into the air as a super volcano. The last such eruption was about 26,000 years ago at what is now Lake Taupo, New Zealand, and there are 47 other known explosions of this magnitude. Researchers estimate that there is a 75% chance that another such explosion will occur in the next million years and only a 1% chance in the next 460–7200 years. Interestingly, the total energy from large volcanic eruptions is greater than that calculated for asteroid impacts, so they may be a worse problem than asteroids [17].

The main problem from these large explosions, global cooling, is the same as from nuclear weapon explosions and collisions with large asteroids. A large amount of particulate material and gas shoots into the upper atmosphere and blocks the warmth of the Sun, leading to global cooling. Destructive, but more common, large volcanoes can cause this effect; Mt. Pinatubo in the Philippines had an eruption in 1991 that was 100 times less powerful than the smallest super volcano. Still, this eruption decreased the temperature of Earth by about a degree Fahrenheit and caused a global decrease in ozone in the upper atmosphere, allowing UV radiation levels to increase measurably. However, none of these effects was strong enough to cause a major negative influence on global human society or most organisms on Earth.

The effects of a super volcano would be devastating to agriculture throughout the word and totally disrupt the global economy [18]. The chance of a super volcano explosion with global effects is greater than that of an asteroid explosion of the same magnitude. Unfortunately, there is little humanity can do about such explosions. The inability to do anything to stop an eruption, and the low probability of an eruption occurring, leads me not to discuss super volcano eruptions much further in this book.

Gamma Ray Burst, Red Giants, and Other Global Catastrophes

There are several scientific observations that mean, ultimately, all human life will become extinct. The Sun will evolve and change, eventually becoming a red giant. At that point, the Earth might either fall into the Sun's atmosphere because tidal drag would slow our orbit rate, or Earth would orbit immediately outside the Sun's atmosphere, and the intense heat would cook all water and life off the surface [19]. The Sun will undergo this process about 5 billion years from now.

An alternative doomsday scenario could unfold with the evolution of stars. Regularly, somewhere in the universe, a massive star will convert to a black hole after becoming a super nova. This process releases an enormous burst of gamma rays along the axis of rotation of the star. If such an event occurred within our galaxy and directly hit Earth, all life would be extinct.

The closer we are to the event, the more energy we would receive and the longer the event might be. Planets on the margins of the solar system are safer than those in the interior (those closer to more stars). The probability is that through the evolution of the universe, most planets in a solar system will be subject to a gamma burst at some time during their existence, and possibly such a burst accounts for one of the global extinction events on Earth over the last 3 billion years [20]. Since the probability of such an event is extremely low, and we likely could do nothing about it (except general civil preparedness), I don't consider destruction by gamma ray in much detail.

Assuming humanity escapes a gamma burst or plunging into the Sun, there still will be a time, hundreds of billions of years from now, when the heat death of the Universe occurs. Eventually all energy that was created with the Big Bang at the start of the Universe will be lost to entropy. At this point, no life or motion would be possible. Alternatively, if the expansion of the universe reverses (if there is enough mass to pull the Universe back into singularity), all life is impossible in "The Big Crunch." Since humanity can do nothing about either of these outcomes, and they are so far in the future, I do not discuss them any further.

Techno-threats

There are a series of potential threats related to humans "messing" with nature. Probably the most worrisome of these is bioterrorism that seems likely enough and is within current technological capabilities, so I will discuss it directly in the next chapter. Global environmental threats are also a result of current technology, and I consider these in Chap. 7. However, there are additional potential technological threats that are less likely or at least less predictable that we consider here.

One of the more science-fiction type scenarios has artificial intelligence attaining self-determination. Once it is able to reproduce and support itself, the scenario goes; this new form of consciousness no longer needs humans and we are competing with it for resources. The artificial intelligence then wipes out the human race it no longer needs. It is impossible to predict if this will happen and when, although we continue to improve computers and artificial intelligence. Given the uncertainties, I will not discuss this further.

Synthesis and release of chemicals into the environment could cause an additional potential threat. Paul Ehrlich, a respected scientist who has thought long and hard about the future of humanity, laid out the following scenario to me. He asks: what if humanity synthesized a chemical that unknowingly caused sterility in all males after 15 years of exposure at extremely low levels? If we released this compound into the environment in large quantities, we could never get it back out. Such a scenario may be unlikely but not impossible; we now know that many common compounds synthesized by people act as hormone mimics and have biological activities at very low concentrations. For example, ecoestrogens (compounds that mimic estrogens) are commonly found in the environment and are causing feminization (males converting partially or wholly to females) of fish and amphibians in many parts of the world. These compounds include the once widely used pesticide DDT [21]. With syntheses of numerous new chemicals that we release into the environment without testing, this outcome becomes ever more possible. We now produce some novel chemicals, such as nanomaterials, about which we have little knowledge of potential health effects. Prediction of the actual probability of threat from this source is probably impossible, so I will not discuss the general problem further.

High-energy physics provides another potential for worry. When the United States detonated the first atomic weapon, physicists were not 100% certain that a nuclear reaction would not propagate and destroy the world. More recently very high-energy physics experiments cause concern. These experiments accelerate microscopic particles and cause them to smash into each other with extreme velocities creating very high-energy collisions in an attempt to probe their structure and form or test hypotheses about their properties. According to physicist Martin Rees [22], there are three possible disaster scenarios related to colliding high-energy particles. The first is creation of a tiny black hole that would suck in all matter around it. He finds this outcome very unlikely given current knowledge of physics. The second would be that the quarks (the subatomic building blocks of matter in our universe) could compress into a strangelet (a strange, very dense, agglomeration of different types of quarks) that contagiously create a new kind of matter, destroying the type of matter that currently makes up our world and converting Earth into a dense sphere 100 times smaller than our current planet. The third is that the particle collisions could cause a phase transition that rips the fabric of space itself, causing all things in our reality not to exist instantaneously. He finds all three of these unlikely because particle collisions that occur regularly in the Universe that are much more energetic than we are able to create in experiments have not yet caused such outcomes. However, as experiments advance, they are coming closer to exceeding energy levels that occur regularly in our Universe. Calculating the probability that such experiments will cause catastrophic destruction is difficult, but has been done, and I will discuss the probability in Chap. 8.

Religion: The End of Times

The eschatological (end of times) writings or legends of many religions and cultures concern themselves with the end of human history or the ultimate fate of humanity [23]. Christianity, Judaism, Hinduism, Buddhism, and Islamism all have stories of the end of the Earth. Zoroastrian, Hopi, and Lakota legends also refer to the end of time. The end of the Mayan calendar generated considerable interest in the end of time as well (though we seemed to have survived it). Every religion must consider the ultimate fate of humanity, because this is one of the deepest questions people ask. Answering such questions is the province of religion. Those with vengeful gods may need to worry more about an unpleasant end to life on Earth as we know it than those with deities that are more beneficent.

Different cultures have diverse explanations of how and why the world will end. Even within a religion, various sects believe in different outcomes for humanity. Nativist movements have many versions of end of times including people of the Andaman Islands (earthquake destroys the Earth), the Altaic Tartars (the Emperor of Heaven will return at the end of the world and judge all people), and the Salish Natives (the creator will return and happiness will remain). End of days are also prophesized by more widespread religious movements or sects.

In Hinduism, the current Earth ends after a degradation of morals, and a new one begins following an apocalypse. Buddhism's goal is escaping Earth and a cycle of birth and resurrection based on the idea that the world is only an illusion (i.e., the illusion is destroyed). Some Buddhists believe in cycles of death and rebirth, including a time when people will all have swords and hunt each other until only a few people remain.

Islam, Christianity, and Judaism all have versions of an apocalypse with terrible suffering. In general, the belief is that this will be a time of resurrection of the dead. While branches of Christianity vary in their versions of how the end of days will occur, they involve tremendous suffering but ultimate rewards to those who believe in the correct religion and punishment to those who do not.

Sunni Islam and Shia Islam are not traditionally considered messianic religions, but they still have writings that detail the end of days. Potential signs include a beast that will harm people, the Sun rising in the west, caving in of the Earth, smoke, and the devil on Earth. The book of Daniel leads Jews and Christians to believe in apocalyptic events.

From a scientific point of view, none of these visions of apocalypse is amenable to verification. They all differ in when and how the world will end. If you accept documents such as the Koran or the New Testament to be literally true, and exclusive of all others, this might be enough proof for you. If there were ways to determine which of the prophecies is true, we still would have no individual control over when it will occur and the outcome for humanity, so there is little benefit to dwelling on it. Since I am basing this book on verifiable facts, I will not consider the end of days in a religious context any further.

Catastrophic Failure of Societal Order?

Predicting catastrophic failure of societal order is very difficult, and this becomes the most nebulous line of reasoning toward deciding what the worst problems are. Many people believe that some sort of moral or religious breakdown will ultimately lead to society plummeting into disorder (and this is a central tenet of the "end of days" prophesies from various religions). Others see the rise of religious extremism as a major threat. Predicting this type of human behavior is difficult, and these issues certainly deserve serious consideration.

The capacity for rapid failure of societal order is very clear. A hasty descent into lawlessness has occurred in many countries at times. The atrocities of war verify the fact that some "civilized" people will perpetrate unspeakable horrors. When hurricane Katrina destroyed much of New Orleans, the complete failure of law and order was obvious. Some people did horrendous things to others once they realized the lack of consequences for such behavior. When sports teams win or lose in major games, people riot in the stadiums or streets of their home cities leading to complete disorder. Soccer, basketball, baseball, ping-pong, and cricket matches have all resulted in rioting fans. Again, a relatively short lapse in societal control can rapidly lead to anarchy.

Anything that causes people to think they lack security can lead to at least partial breakdown. The global financial crisis of 2008–2009 is a good example. Nothing clearly identifiable changed as far as the ability of Earth to support people. Potential rates of food production did not decrease, and the rates of death from war or disease were not substantially different from other years. Yet, many people suffered because the price of food increased and the flow of money decreased. Human behavior is notoriously difficult to predict, and who would guess that risky banking procedures could result in increased hunger around the world (although it comes as no surprise that the poorest people suffered the most because of the activities of the richest people).

The ability to predict nonlinear behavior of complex systems is limited even for the simplest systems, and human behavior is not simple. The idea that uncomplicated systems can undergo rapid and drastic change has received much study, but we have little ability to predict when that rapid change will occur. The classic example is a pile of sand. If we add sand to the top of the pile in a very regular fashion, the pile will continue growing to a point. Eventually the sides of the pile will collapse outward. It is not possible to predict exactly when those sand-slides will occur. Much of current study of the natural world (ecology) is concerned with understanding collapse of existing systems related to relatively small perturbations. Ecologists are not very good at predicting exactly when an ecological system will collapse, and social scientists are not very good at predicting when societies will collapse.

The metaphor of a house of cards is apropos here. The structure can be flimsy in general, but nobody can predict when the entire house will collapse if cards are carefully added and nobody purposely knocks it down. Human society is similar because nobody can predict when the threat of negative effects of authoritarian control will lead to revolution. The "Arab Spring" caught most people by surprise. People generally cannot forecast when societal breakdown or strong changes will occur. Such is the case in Somalia where there was no central government between the civil war in 1991 and reestablishment of a unified government in 2012. This was one of the poorest and most violent countries and home to vicious pirates and general lawlessness. What exactly leads to this type of collapse and is a similar collapse possible globally?

Some problems will clearly result in global collapse, and this is easier to predict. Anything that will disrupt the ability to provide food to most people on Earth will cause societal breakdown. Possible scenarios include global cooling from nuclear winter, global death of crops from increased ultraviolet radiation, and continuously increased demand for food when science and society are unable to produce more food from existing cropland to feed an expanding global population.

Additionally, breakdown of technology (e.g., nuclear war leading to failure of most electronic devices), ability to transport goods, decreased energy availability, and any other widespread disruption that would interfere with global economic systems could have disastrous consequences for millions or even billions of people.

Situations that could lead to mass panic and societal collapse include, but are not limited to, global war, spread of disease that kills or debilitates a substantial portion of the world's population, and widespread panic related to economic or political collapse. Where possible, I will mention where the potential exists for such catastrophic collapse. In the next few chapters, I will go through current causes of death and suffering, and the instances most likely to lead to death and suffering in the future because in large part they are likely to trigger social collapse. I will cover this topic again in the last chapter.

Chapter 4
Disease Now and Potential Future Pandemics

The plague of Florence in 1348, as described in Boccaccio's Decameron

A group of wealthy travelers enters your village, a day's ride outside of Florence, Italy, in 1348. They offer a handsome sum for housing and food, and are fed and given rooms. Later that evening one of them is overheard talking about a terrible disease that has started in Florence, expressing relief they have escaped. The next day, one of the travelers calls for a healer. He has pustules covering his body. Within 4 days of getting a high fever and vomiting blood, he has died. Soon others in his party and then people in the village start exhibiting the same symptoms. Within 2 months, 3/4 of the people in the village are dead. A few attempt to escape and travel to other villages. The Black Death (bubonic plague) has taken another village and continued to spread. This pandemic was the worst in recorded history and

© Springer Nature Switzerland AG 2019
W. Dodds, *The World's Worst Problems*,
https://doi.org/10.1007/978-3-030-30410-2_4

caused the death of roughly half of Europe's population. It killed one in five people on Earth, and caused drastic political changes [24].

Right now, a child lies weak in a hut in Africa. She has not been able to hold food for a week and is nothing but skin and bones. Only a moment later, the life passes from her eyes, another victim of diarrhea. This will happen again to four children under the age of 5 somewhere in the world in the next minute or two.

These two stories illustrate two huge problems related to disease. First, I discuss disease as a major ongoing cause of mortality in developing countries, particularly for children. Next, the potential for a pandemic disease that kills a large portion of the population is covered. As we shall see, a massive pandemic is becoming ever more probable based on predictions arising from simple scientific principles. I start the chapter with some basic biological information because this information is necessary to understand how diseases develop and how humanity might be able to control them. Once again, taking the bias out of confirmation with increased understanding of accurate scientific information improves clarity with respect to causes and solutions of problems.

What Causes the Deadliest Diseases?

Diseases have probably existed since soon after the first life evolved. All organisms known today have diseases. These infections harm bacteria, fungi, protozoa, plants, and animals. Many of our diseases have their own diseases. Disease is an inescapable feature of human life. Humans can contract communicable diseases from bacteria, fungi, helminthes, protozoa, viruses, and prions. A little biology, my favorite subject, is required to understand how they cause disease, how novel diseases can arise, and why some of them are so difficult to control.

Helminthes are simple animals, little microscopic worms. These animals often lack digestive systems, as they take nutrition directly from their hosts. They have complex life cycles and can inhabit several hosts. Evolution has stripped these animals down to only the basics required for living off others. Schistosomiasis is the most common helminth disease, causing widespread suffering. Initial infection can cause fever, chills, and liver and spleen enlargement. Chronic infections can lead to liver or kidney failure, bladder cancer, heart failure, and other symptoms. This disease is difficult to treat and causes major suffering in the world; at any one time, over 200 million people are infected, mostly in Africa. This disease is a major cause of suffering, but we don't know enough to eradicate it. We have studied most tropical diseases far less than those influencing populations in wealthier temperate countries. Worldwide cooperation in study of all major diseases is required to decrease the amount of suffering related to helminth diseases.

Fungal diseases are becoming more dominant as sources of mortality. They are particularly problematic for immune-compromised patients, but some of the diseases are endemic in environments around the world. They cause less mortality than many other diseases.

Protozoa are single-celled microbes that are tremendously diverse and play both positive and negative roles for humans. Protozoa cause a variety of nasty diseases including African sleeping sickness, *Giardia* and *Cryptosporidium* (two major causes of diarrhea). These diseases are difficult to control because protozoa are more like us than bacteria and chemicals that kill them tend to harm humans.

Of the diseases caused by protozoa, malaria is one of the top diseases causing death and suffering in the world. Mosquitoes carry this disease, and it is common in tropical areas worldwide. As global warming increases temperatures, mosquitoes will move into formerly colder climates, spreading malaria to new areas. There are over 200 million cases and a half million deaths each year from malaria. This is one of the major scourges of humanity, and it is a difficult disease to prevent and cure. More research is essential to control this disease.

Prions are strange disease agents that arise from proteins that cause other proteins to change shape and become ineffective. The mad cow disease is probably the best known of these diseases. This prion-caused disease arose when meat producers in Europe fed animals neural material (brain and spinal tissues) from other cattle. Cannibalism is not natural for cattle, but the practice arose because it stimulated growth when animals consumed waste by-products from other slaughtered cattle. Some of the sick animals had a protein in their brains that was defective and could infect the brains of other cattle when ingested. If people unknowingly eat beef contaminated with some of this infected neural tissue, they also become infected. The defective protein causes human neural proteins to lose their function slowly over time. People have died from this and related prion diseases. However, this number is small relative to other diseases and I will not consider it further.

Some of the worst diseases of humankind are bacterial, including the Black Death (plague), tuberculosis, cholera, salmonella, and coliform diseases. While antibiotics allow treatments for many of these diseases, they still infect and kill millions of people each year. Bacteria are very simple cells and can completely replicate themselves if provided the basic vitamins and compounds they require. The vast majority of bacteria are harmless to humans and even beneficial. For instance, if we did not have bacteria that decompose dead animals and plants, the Earth would rapidly fill with these and there would be no resources left for the living.

A very small number of bacteria, relative to the total millions of species on Earth, can cause significant disease in humans. Of these, some do so incidentally and can live indefinitely in the environment without ever interacting with a human in any way. However, a few can only live in animals and can lead to debilitating if not fatal diseases. Intermediate between these two strategies are the bacteria that opportunistically cause disease in humans, but can live freely without an animal host.

Viruses, like prions, are not individual life forms. They are more like information parasites that completely rely on their hosts for reproduction. Viruses take advantage of the fact that we use DNA as the blueprint containing the information we use to make all cellular components and RNA to translate that information from DNA into the structure and machinery of the cells (proteins). Viruses subvert the cell's reproductive machinery for their own nefarious purposes: the reproduction and dispersal of more viruses. All viruses, unlike bacteria, must take advantage of living cells with DNA to reproduce. Viral diseases include HIV, smallpox, polio, hepatitis, influenza, Ebola, and rabies. Viral diseases also infect and kill millions of people per year.

The Biological Arms Race: Disease as an Evolutionary Process

Our bodies and our species are engaged in ongoing arms races against our diseases, just as all other species are in a biological arms race with their diseases. In the short-term, our immune systems learn to recognize and fight diseases. In the long-term, humans have evolved defenses against the diseases. For each defense humanity has evolved, the diseases have evolved ways to get around the defense.

The Red Queen Hypothesis describes this arms race. The Red Queen from Lewis Carrol's *Alice in Wonderland* needs to run as fast as she can just to stay in place. This hypothesis suggests that organisms need to evolve as fast as they can to keep up with the organisms with which they interact because they are evolving as well. This is a case of feedback, as both members of the interaction need to keep evolving to survive.

In the case of human diseases, this arms race leads to a situation where humans are constantly adapting to their diseases, and the diseases are constantly evolving to take advantage of their hosts more effectively. To make things more complex, a disease that can jump from host to host, such as the flu that can move from pigs to humans, can escape the evolved defenses of one host by moving into another. We will consider this more completely when considering diseases arising from other animals. Sometimes these crossover diseases are too successful; in this case, the disease can completely kill off its new host leading to extinction of the disease itself. It is certainly a worst-case scenario that such a disease would arise in humans.

Humans have an immune system that can protect against most invasive diseases, but once an invader evolves a mechanism to skirt the human immune system, it can infect people everywhere. Infections will occur as long as humans do not evolve a way to keep the disease from reproducing or find a way to avoid exposure to the disease. Evolution to evade human immune defense is why viruses that cause colds move though human populations so readily. The group of cold viruses has evolved a way to swap genetic information, continuously creating new viruses. There are around 100 strains of cold virus, and each person gets a cold roughly twice a year. Thus, it would take you 50 years to go through all the cold viruses… and in that time new viruses could evolve.

The process of evolution is not "nice," because the very basis of natural selection is that some organisms die and others that are better adapted survive. When a novel deadly disease arises, it may not kill all the people that it infects, but could kill most and only the lucky few who survive can reproduce and pass their offspring resistance to the disease. Likewise, if humans become very resistant to a disease either through evolution or by technology (e.g., an effective vaccine), it might die out.

The ideas of humans evolving resistance to diseases, and of diseases becoming more effective at infecting people, lead to some of the major points in Jared Diamond's book *Guns, Germs, and Steel* [25]. He suggests that successful invading cultures are those that have evolved to be resistant to a larger array of diseases and as they evolve this resistance, the diseases become more virulent (infective) in order to spread. This evolution can lead to a situation where people are no longer killed by a disease, and some of them become disease carriers. Such a group of people will bring these diseases with them upon colonizing a new area. The disease that is not lethal to the colonists could be quite lethal to the colonized people that have no evolutionary experience with it.

European colonization of the Americas is a prime example of this form of migration and conquering other groups of people. Agriculture and technology made possible the large populations and cities in Europe and Asia. People living in dense urban environments, and trade routes connecting those groups of people across vast areas, created the perfect breeding grounds for new diseases to develop. Living in close proximity to domestic animals further stimulated the emergence of new diseases. The huge land mass of these two connected continents (and a modest connection to Africa as well) meant that there were many locations for diseases to develop then spread across the continents. Exposure to repeated waves of diseases allowed Europeans to develop immunity to those diseases while still chronically harboring many of them.

When Europeans encountered the indigenous peoples of the Americas, they gave them their diseases. These diseases probably moved across the American continents faster than the Europeans invaders did. The resulting wave of diseases could have killed even more people than the Black Death of Europe. We will never know exactly, because there are no written records and no accurate methods to estimate the exact causes of death of the pre-European human population in the Americas. However, archeologists have found evidence for a large crash in American Indian populations across both North and South America after first contact with Europeans; the estimates are 57% mortality. These estimates are based on genetic methods that can be used to indicate drastic population reductions (from all causes) in the past [26]. Thus, the European colonists entered a land with relatively low population density that was poorly defended with a huge amount of exploitable resources.

The endless cycle of disease, massive death, development of resistance, and resurgence of disease was the hallmark of human history until we developed the technology to control many of our diseases. Cultural evolution moved ahead of biological evolution and protected us by changing our behaviors and ultimately by allowing us to develop the tools to manipulate the biochemistry of our own cells to fight off diseases (inoculation with vaccines).

While many animals have adaptations that help them avoid disease, one of the earliest unique signs of human cultural technological developments to avoid disease (and make food more digestible) is the act of cooking food. Additional steps to avoid diseases included cultural changes such as dietary restrictions of religions, as well as guidance in matters of personal hygiene in ancient cultures and religions.

By the mid-1800s, John Snow used statistical methods to link a cholera outbreak in London to a particular well on Broad Street. The authorities shut down the well, probably helping end the outbreak. This understanding of the source of disease was the first case of using epidemiology (the science of incidence, distribution, and possible control of diseases) to help understand how to control disease [27].

At about this time, Louis Pasteur developed the idea that bacteria spread disease. Once people had microscopes, they could see bacteria and assign them as potentially causative agents of disease. These observations by Pasteur and his contemporaries led to a rapid succession of ways to control diseases including purification of drinking water, preservation of food (pasteurization), and sterilization or sanitization during surgeries.

It was not until the discovery of antibiotics and vaccinations that humans really started to free themselves of some of the worst diseases in history, at least temporarily. People started looking for chemical agents to destroy bacteria once science confirmed that they caused diseases. Alexander Fleming discovered penicillin in 1928, and in 1942, Howard Florey and Ernst Chain developed the drug penicillin in an easily produced and administered form. Yet, humans still did not win the evolutionary battle between bacterial disease and humankind. When Fleming gave his Nobel acceptance speech in 1945, he noted that bacteria could develop that were resistant to penicillin if exposed to less than lethal concentrations. Within 2 years of the advent of clinical use of penicillin, clinicians had noted antibiotic-resistant bacterial infections in human patients.

Antibiotic resistance is a classic case of evolution. A very low proportion of bacteria in a population have the ability to survive exposure to the antibiotic. There is a low level of mutation leading to very small differences in the DNA in each bacterium. By chance, one or a few have a mutation that allows them to escape death from the antibiotic. These bacteria reproduce and soon take over the entire population.

Bacteria are particularly well suited to rapid evolution because their populations are so huge and grow so quickly that the chance that one cell has a mutation that makes it resistant is high, even though the chance that each individual cell is resistant is extremely low. There are about 100,000,000,000,000 (a hundred trillion) bacteria associated with each human. This is about 1000 times as many as there are stars in our galaxy, or 100,000 times as many as all the people on Earth.

There is an even more interesting, and insidious twist to this story. Bacteria have special little bits of DNA called plasmids. Since bacteria are not sexual organisms, they cannot exchange genetic material through sex like we, other animals, and many plants do. Instead, they trade these little plasmids. They can trade this genetic material within or among species. Thus, antibiotic resistance can move among different species of bacteria.

A recent spread of antibiotic resistance illustrates the problem [28]. The antibiotic colistin represents a last-ditch compound used to treat infections that are resistant to multiple antibiotics. The gene mcr-1 started showing up in bacterial infections in hospitals around the world, and this gene makes bacterial infections resistant to colistin. The antibiotic colistin had seen limited use in humans because it has bad side effects. However, swine farmers use it. Researchers sequenced isolates of colistin-resistant bacteria from 31 countries, and the genetic data suggested the gene arose in a pig farm in China in 2006. It took less than 12 years for disease-causing bacteria containing the gene to move from livestock to humans globally.

Numerous plasmids for antibiotic resistance have spread readily around the world. Now microbiologists can take a sample from the center of the ocean and isolate bacteria that are resistant to antibiotics only synthesized by humans thousands of miles away. Thus, in the evolutionary battle, we cannot easily vanquish disease-causing bacteria.

Controlling the expansion of resistance to new and existing antibiotics requires an understanding of evolution. Taking a full course of antibiotics to be certain the drug completely clears the infection, only using antibiotics when they are necessary, and not allowing livestock producers to use antibiotics added to feed solely to increase growth rates of healthy animals are all necessary to decrease the probability that bacteria will become resistant to antibiotics. Cooperation not to overuse antibiotics, leading to antibiotic resistance, is necessary to control this threat. This cooperation needs to be worldwide given the propensity of diseases and antibiotic resistance to spread rapidly around the world. This is an evolutionary "arms" race, with bacteria rapidly evolving ways to resist antibiotics and humans developing new weapons against bacteria. We need science to win the "arms race" by coming up with novel antibiotics to stop diseases resistant to the current antibiotics. Otherwise, we end up back where humanity has been for most of its history, at the mercy of bacterial infection.

Chronic Diseases of Human Kind: Why Do So Many Children Die of Preventable Diseases?

We live in a world where millions of people die each year from preventable diseases. These diseases often only cause temporary illness to people in developed countries. For example, in developed countries, we do not generally consider diarrhea a fatal disease, and we successfully treat most severe cases with hydration and chemical therapy. In bad cases, intravenous fluids can prevent dehydration. Likewise, many cases of pneumonia can be treated, particularly bacterial-caused cases that are treatable with antibiotics. However, pneumonia is the top cause of death, diarrhea the second, and malaria and problems with childbirth are the next two top killers of children when considered worldwide.

Diarrhea causes 15% of global deaths in children under 5 years of age even though it is treatable with access to basic medical care and many cases are preventable with clean food and water. Pneumonia causes even more childhood deaths (18%). Essentially all of these deaths occur in developing countries. Vaccines against some bacteria that cause pneumonia can help, and access to antibiotics for bacterial cases can help cure the disease. Inoculations against measles and whooping cough (pertussis) can also help reduce mortality as these diseases can lead to pneumonia. Malnourished children are more susceptible as well as those are that are exposed to indoor air pollution (smoke from cooking fires). Only about 1/5 of the children with bacterial pneumonia even have access to antibiotics. Solutions are obvious and not tremendously expensive. Still the solutions to these diseases require global cooperation and a concerted effort.

There is good news; child mortality rates continue to fall, and diseases are decreasing worldwide [29]. However, in undeveloped countries, rates are still high and not declining rapidly enough. The following interventions could reduce childhood disease (in parentheses) by over 40%: (1) breastfeeding (diarrhea, pneumonia, neonatal sepsis), (2) insecticide-treated nets (malaria), (3) additional food for infants 6–12 months old (diarrhea, pneumonia, measles, malaria), (4) zinc supplements (diarrhea, pneumonia), (5) sanitary delivery (childbirth deaths), (6) Hib vaccine (pneumonia), and (7) clean water (diarrhea, pneumonia). Oral rehydration therapy, antibiotics, and antimalarial therapy could reduce deaths by another 31%. None of these steps seems that difficult.

Globally, including adults and children, the World Health Organization estimates that millions of people are afflicted by treatable diseases. These include (in millions), diarrheal disease (4620), lower respiratory infections (429), malaria (241), measles (27), pertussis (18), dengue (9), tuberculosis (8), and HIV (2.8). Many of these diseases are disproportionately harming people in less developed countries.

Spread of Successful Diseases: Why Epidemiology Matters

The study of epidemics is essential to understanding how diseases spread from one organism to another. Diseases can be passed from person to person, from animals to people, or from the environment to people. A firm set of general principles of the conditions under which harmful diseases will emerge, some of them already referred to here, are now widely agreed upon by epidemiologists. These general principles, in addition to a number of unique behaviors in humans, lead to the conclusion that the probability of global disease is increasing drastically.

Density of hosts is important in proliferation of disease. In diseases that move from person to person, the greater the frequency of contact between people, the more quickly disease can increase. We now live in a world where over half the people live in cities. The total number of people on Earth is still increasing, leading to ever-greater population density.

People move from city to city and around the globe more quickly and more frequently than ever. Many diseases have an incubation time of days to a week. Historically, long-distance travel across oceans took months. If somebody with a deadly disease boarded a boat, it could kill all the people on the boat, or they would be over the disease and no longer infective before they reached their destination.

Thus, conditions are more conducive than ever for the spread of a new deadly disease around the world.

New diseases are more likely to arise when there is opportunity for them to take hold. Stressed or unhealthy people are more likely to contract and transmit diseases than healthy individuals. We now have a greater absolute number of disease-prone malnourished people on Earth than at any time in history. People who live in cities, particularly the poor, are under constant stress from crowded living conditions, unhealthy environments, and the pressure to survive. A greater proportion of people live in cities now than any time in history, and the total number of people living in cities is also greater.

Novel diseases arise when humans come in contact with animals that can transmit diseases to humans (zoonotic diseases) [30]. We already have dozens of diseases that come from animals. We grow ever more animals to feed people as increasing population and standard of living increases demand for meat. We grow these animals under crowded conditions. Oftentimes we breed the animals for maximum growth, and they are genetically similar, conditions that can lead to greater disease susceptibility. We raise many pigs and birds in huge numbers in close proximity. Most diseases that arise in birds are not transmissible to people. However, they are more transmissible to pigs. Once the bird diseases establish in pigs, they are more likely to evolve ways to be infectious to humans.

Bush meat refers to wild animals killed by people for protein. As people move into more remote areas, they hunt wildlife to satisfy their protein needs. There is a particularly high chance of disease transmission from animals to humans when people butcher wild animals and make contact with fresh blood and organs from these animals. Human populations are growing and moving into more wild areas, particularly in the tropics.

Most new human diseases come from wild animals. As people stress natural ecosystems, the probability for such transmission could increase because we displace other animals from their natural habitats causing them to move longer distances, and they are more susceptible to disease because of the stress induced by human pressure. There is also a demand for live exotic meat in some Asian countries in particular. In this case, merchants bring caged animals into dense markets in close contact with other animals and many people. These are perfect conditions for a novel disease to spread from infected animals to livestock and people.

More people in the world are now immune compromised than ever, offering another reservoir for diseases. Around 40 million people are HIV positive and may become immune compromised. In areas where medical care is available, these people are receiving quite a few antibiotics leading to increased antibiotic resistance evolving in bacterial diseases. In addition, recipients of transplanted organs are generally immune compromised because drugs given to them to avoid rejection of the transplants also halt immunity to disease.

Humans have also developed some novel behaviors that further increase the probability of transmission of diseases, particularly those that require blood-to-blood contact. Intravenous use of illegal drugs leads to sharing of needles and has resulted in transmission of HIV and Hepatitis B and C. This behavior could also lead to transmission of novel diseases in the future. Blood transfusions and organ transplants can also lead to movement of diseases. People receiving blood products were some of the first victims of the HIV pandemic.

Organ transplants from other species into humans could also lead to transmission of novel diseases. The most common species considered for organ transplants into humans are apes and pigs. Both these groups of animals carry viruses that may be harmless to them but cause extreme disease in humans. Such transplants are very rare currently, but given that the majority of people on organ transplant lists are not able to obtain an organ before they die, it is likely such transplants will increase as research progresses on ways to accomplish such transplants successfully.

The field of epidemiology tells us that it is becoming more probable that new diseases will develop, existing diseases could evolve into more deadly and virulent forms, all diseases will multiply more quickly, and they will continue to become resistant to human efforts to control them. These findings have ramifications for both chronic diseases and new diseases.

Emergent Pandemics and Superbugs

One of the worst worldwide pandemics was the "Spanish" flu that started in 1918. It killed about 3% of the world population and infected about 1/6 of all people. The bubonic plague in the 1300s infected about 1/4 of the Earth's human population and killed an estimated 13%. The "swine" flu (H1N1) started in 2009 and infected about 1/4 of humanity but killed less than a hundredth of 1% of our population. Scientists have traced the first widespread series of cases of HIV/AIDS to 1981, but the disease probably jumped into humans in the early 1900s. Since then, about 1% of people on Earth are living with HIV, and about 1.5 million people per year die because of AIDS. About 2% of the human population deaths each year is from AIDS-related causes worldwide. Waves of disease are a regular occurrence throughout human history and becoming more common.

Recently the world has been concerned (terrified) about Ebola. Symptoms include fever, severe headache, muscle pain, weakness, fatigue, diarrhea, vomiting, abdominal (stomach) pain, unexplained hemorrhage (bleeding or bruising), and death. This disease has been simmering in Africa for decades. Outbreaks have occurred in sub-Saharan Africa regularly since 1976; in 2014, an outbreak started in Guinea and jumped to other African countries in weeks and then to countries around the world killing over 10,000 people. In 2019 almost 2000 people died in the Democratic Republic of the Congo, and stopping the disease there has been difficult because of warfare; this outbreak has spread to Uganda. The ease of global movement and increased travel continue to increase the potential for spread of the disease. What if this disease evolves to an even more easily transmitted form? There is no treatment or vaccine (although some promising vaccines are being developed).

Disease epidemics that do not kill a large proportion of the human population are common. In the 1700s there were 13 epidemics and in the 1800s 12, with 5 pandemic influenza epidemics in the 1900s. The data suggest that roughly every 10–20 years, there are epidemics with some mortality that infect a quarter to a third of the world's population.

You could argue that disease has not wiped out humans yet, so it will not in the future. Unfortunately, science has documented cases where diseases cause the extinction of an entire species. For example, people have inadvertently moved a fungal disease around the world that kills amphibians (frogs and salamanders). This disease is leading to numerous species extinctions globally. I have seen the effects of this disease first hand in Panama.

We studied the consequences of the fungal disease killing all adult frogs leading to loss of all the tadpoles in Panamanian mountain streams. Scientists had already documented that the disease was moving through North America to South America through Central America. The disease kills frogs in high-elevation areas and moves through lower-elevation areas without killing most animals. We knew that the area we were working in was going in the direct path of the disease, so we set up a before-after experiment to understand the effects.

On our first visit to the mountain stream, there were frogs everywhere. We needed to be careful not to step on them as we walked the trails. Each square yard of stream bottom had up to a hundred tadpoles. Twenty frog species used the streams for reproduction, and many of these species were entirely restricted to cooler areas with high altitude on this particular volcanic mountain. From sunset to sunrise, the jungle was a riot of frog choruses. There were fantastically colored adult frogs including the stunning black and white Panamanian Golden Frog, a species that has special meaning for Panamanians. We made our measurements on the stream, and enjoyed the frogs.

Two years later the disease had swept through as it progressed through the country from Costa Rica. When we drove up to the stream for the "after disease infection" experiment, it was immediately clear that it was different. Hoping against hope, I went down to the stream, but there were no tadpoles and no adult frogs on the banks. It was very quiet and sad. The stream had dense growths of algae because no tadpoles were eating it and the absence of the tadpoles fundamentally changed the way the stream functioned. In the end, maybe 100 species will go extinct from this disease.

Through this and other examples, we know that some diseases have driven animals and plants to extinction [31]. In Hawaii, 16 cases of bird extinctions have come about at least in part because of diseases. Numerous mammals and birds have gone extinct from diseases alone or in combination with other factors such as habitat loss [32]. Thus, it is not impossible that humans could suffer the same fate. The conditions that could lead to such a tragedy are making it more likely that such a disease could infect the human population.

Throughout human history, new nasty diseases have arisen. Many of them have jumped into humans from other species. Whenever a particularly virulent disease has infected a human, and killed most of the people exposed to it, the population of people infected was small and disconnected from the rest of humanity. Epidemiology tells us that the incidents that were formerly isolated now have the potential to sweep across the globe and cause massive death and suffering.

We are increasing the conditions under which such diseases can arise and transmit to people (ever more intimate contact with wildlife, dense livestock production). Losses of biodiversity caused by humans also are predicted to increase the transmission of infectious human diseases [33]. It is no wonder that new diseases like Avian flu, H1N1, Ebola, and SARS are popping up with alarming regularity. Adding to the worry, viral evolution is unpredictable, and a new deadly strain of virus could arise from laboratories working on viruses that are presumably safe and contained. In this case a newly virulent strain could arise, escape, and become a pandemic [34].

At the same time, new diseases challenge the safety of people and the ability to treat such diseases increases. We can develop vaccines rapidly. Antiviral drugs are available that work at least well enough to decrease mortality. However, only those people in developed countries are able to afford or even have access to methods to protect from sickness and death from these infections. As usual, the poorer people of the world will suffer the worst of globalization, increased population, greater chance of new diseases, and unequal distribution of basic health care.

Bioterrorism, Biological Warfare, and Accidents

In late 2011 and early 2012, two laboratories, one at the University of Wisconsin-Madison, USA, and the other at Erasmus MC in Rotterdam, the Netherlands, found out how to make avian flu (H5N1) transmissible in ferrets. This research ignited a firestorm of controversy because the deadly virus could also spread among humans much more easily. The researchers submitted the work for publication but journals held up the release of the papers because of fears that people with bad intent (bioterrorists or countries willing to employ biological warfare) could use the information to transform this and other viruses to more deadly forms. Ultimately, the journals published the work, as eventually the information would get out. This is the way science works, once the general concept for an important idea is out, somebody else is certain to replicate the experiment. Thus, information on how to create a deadly disease is ever more available.

Accidental release from existing research facilities is also a concern. The deadliest diseases known to humanity are stored and researched in containment facilities found around the world. Smallpox has killed people for at least 3000 years, and following vaccination, it was completely eradicated from human populations in the 1970s. A number of laboratories still keep cultures. In 1978, one person died from exposure to the virus in a British laboratory. After that, scientists transferred all cultures to two laboratories, one in Russia and one in the United States. Entire generations have reached adulthood with no exposure to the disease; if smallpox was ever released by accident or on purpose (a scientist with PhD-level training could potentially re-create it from the known genetic sequence), it could cause massive mortality.

In 1979, the Sverdlovsk military facility accidentally released anthrax causing 100 human deaths. Soviet researchers probably isolated this highly virulent strain of anthrax from rodents in the Soviet city of Kirov. The facility had likely accidentally released the bacterium at least once previously. Anthrax is able to survive as dry spores, and Soviets were presumably producing it to arm biological weapons.

While research on diseases is necessary to learn about causes and cures of the diseases that influence humans, such research comes with a cost. The ability to contain these diseases in research settings is plagued with the problem of potential human error. In addition, the possibility of terrorist attacks on such facilities is perhaps remote, but real.

In 1984, followers of Bhagwan Shree Rajneesh in Oregon released salmonella into 10 restaurant salad bars sickening 751 people in an attempt to keep them from voting in a local election in which the cult had candidates. Luckily, nobody died in this incident, but it does illustrate that people can be capable of bioterrorism.

In June of 1993, members of the Aum Shinrikyo cult sprayed anthrax from the top of an eight-story building in the heart of Tokyo. Fortunately, the disease did not take hold. The strain they used was not very deadly, and they had problems with a sprayer so the dispersion of the disease was not as effective as they had hoped. This group had previously set up multiple laboratories and had experimented with the toxin for botulism, cholera, and Q fever (a dangerous bacterial disease carried by livestock). They also previously sponsored a trip to the Democratic Republic of Congo that was an attempt to bring back an isolate of Ebola. This apocalyptic cult eventually released the chemical weapon Sarin in the Tokyo subway killing 12 people and sickening thousands.

While both these examples are unusual cases, we are entering a world where a few crazy people or one crazy country could do tremendous damage to humanity if they had access to the right materials and knowledge. Such knowledge is becoming commonplace. Every year academia cranks out numerous PhDs around the world with the technical expertise to build a deadly virus with the right equipment, chemicals (reagents), and knowledge of the sequence. At the same time, technology to work with DNA sequences is getting cheaper, easier to use, and more broadly available. With a million dollars and proper training, it is now possible to create designer diseases.

We should consider motives in this discussion as well. A terrorist who wanted to kill many people but wanted to discriminate victims would not only need to create a disease but also vaccinate or protect all the people they did not want to die. While a few doses of a disease placed appropriately could quickly spread around the world, creating many doses of vaccine is a far more daunting and expensive task. Thus, it seems unlikely that any of the major terrorist groups would be able to create a disease and vaccinate large numbers of people before releasing the disease without being detected first. Such a task is not completely out of the question for a small country such as North Korea.

There are insane people who just might try to take down the entire human race. The mass shooting in a movie theater in Denver in 2012 was carried out by a neuroscience PhD student. This individual could have had the technical ability to create a novel disease. A scenario where such a person creates and releases a deadly virus is conceivable. Quite a bit of preparation and disaster training would be necessary to stop transmission of an infectious agent once it was released [35].

How Can We Stop Pandemics?

Active surveillance of disease outbreaks is necessary to react to a pandemic in time. Currently the World Health Organization keeps track of disease outbreaks, but we need new methods to detect pandemics in time to respond. An effort called Global Viral [30] is also helping move the international community toward predicting viral outbreaks by promoting science and education about viral outbreaks.

Responses could include rapid development of vaccines to protect a population, antiviral drugs to decrease the probability of mortality, and increased capacity to produce these treatments. Public health officials are developing novel methods using social media and networks of cell phones to create an early warning system of disease outbreaks from remote areas. Additional improvements are possible in safety of livestock production, safer practices in killing and butchering wild animals, and education on behaviors that discourage the transmission of diseases.

There are several things that we can do to thwart bioterrorism. The easiest way to obtain a disease would be for a terrorist to attack one of highest-level containment facilities in the world that has cultures of the deadliest diseases known to humankind. These laboratories should be resistant to attack and have established procedures to quickly destroy all the diseases in them if they are compromised.

Several historic agreements are present to stop the use of biological and chemical warfare and terrorism. The 1925 Geneva Protocol, the 1972 Biological Weapons Convention, the 1976 Environmental Modification Convention, and the 1993 Chemical Convention all strove to have all countries agree to not develop or use chemical or biological weapons [36]. The Obama administration was involved in international efforts to stop the spread of biological weapons to terrorists. The United Nations is developing other treaties. Measures might include careful accounting of the reagents that molecular biologists can use to produce diseases and barring companies from providing DNA or RNA synthesis for particular sequences known to be associated with disease organisms. Careful international monitoring of biological laboratories capable of handling pathogenic microbes is also necessary, as well as protecting the laboratories that hold the deadliest human diseases. All of these require modest cost, but a substantial degree of international cooperation. For example, if even one country allows synthesis of dangerous virus sequences and will sell those to individuals, control elsewhere is pointless.

Chapter 5
Hunger

Depiction of victims of the Irish Great Famine, 1845–1849 Illustrated London News, December 22, 1849

Famine has been a part of human history since we became a species. The first recorded famine was in Rome in 441 BC. Probably the worst in terms of total deaths was the Great Chinese Famine from 1959 to 1961 where 15–30 million people died. This famine is instructive because it occurred as a result of a combination of factors, from adverse weather to poor environmental management decisions, and includes harmful government policies. Poor management decisions

© Springer Nature Switzerland AG 2019
W. Dodds, *The World's Worst Problems*,
https://doi.org/10.1007/978-3-030-30410-2_5

included deep plowing that was thought to allow deeper rooting, but brought up poor soils to the surface. An effort was made to kill all the sparrows that were thought to be eating the grains; this led to an explosion of the insect pests that attacked the crops. Thus, this famine had many of the ingredients that play into understanding global hunger.

Those living in developed countries (except for the poor and disenfranchised) have probably never experienced hunger, so may not understand what it means to suffer from hunger as many people in the world do. Malnutrition causes wasting of muscles, bleeding gums, dull sparse hair, poor wound healing, developmental neurological delay, lethargic and apathetic behavior, and other symptoms. Cognitive impairment from protein-calorie malnutrition is widespread, as is iron and iodine deficiency. Ultimately, starvation causes the body to waste away as tissues are broken down to sustain vital heart and nervous system functions.

A Billion Hungry People

Of every eight people in the world, one does not have enough food. There were 963 million undernourished people in 2009 according to the Food and Agriculture Organization of the United Nations [37], out of a global population of 6.7 billion. Most (but not all) of these people are in developing countries. The United Nations World Food Programme claims that hunger is the number one risk to human health worldwide.

Hunger is suffering. Those of us in the developed world that are not at the very bottom of the economic ladder rarely experience true hunger. Malnutrition does not allow proper physical and mental development of growing children. Undernourished people are less likely to be productive members of society.

We currently grow enough food to feed all the people on Earth. Hunger is a matter of distribution of wealth and access to food. As the population grows and demand for food increases with increased standards of living, it is less clear if there will be enough food. For now, distribution and poverty are key problems. Projections are for the Earth's population to keep increasing over the next century to between 9.6 and 12.3 billion people by 2100 [38].

The global economic downturn of 2007–2008 combined with increased demand for grain supply for nonfood uses had tremendous implications for high food prices. A US tax subsidy for ethanol production from corn led to at least a doubling of the price for grains over the following decade, according to the Food and Agriculture Organization. The price of meat remained steady, but the price of food oils almost tripled. Since the lowest-income people rely on grains as their main source of calories, and a very large portion of their income goes to food, the increase in food prices has hit the people hardest that are closest to undernourishment.

How Will We Feed a Growing Population?

Food production is increasing at a faster rate than human population is growing. However, demand is growing more rapidly as people in developing countries consume more animal-based (meat and dairy) diets [39]. Producing animal-based diets generally requires more crops than vegetable-based diets, so demand for food is skyrocketing. Are there vast untapped sources of food we can exploit to allow all people on Earth to have healthy diets now and into the future?

In the past, some have claimed oceans can feed humanity. They are vast and in some areas very productive, particularly in areas close to the coasts. This high productivity in areas where people observe the oceans leads to a somewhat incorrect concept that the oceans have huge untapped potential. Most of the ocean is dominated by nutrient-poor areas with productivity levels similar to the deserts of the world. Humanity has overfished most of the major fisheries, with substantial decreases in the populations of the large fish that people like to eat, so total amounts harvested are already declining [40].

Still, the oceans have great potential for feeding a large human population. We need to change the way we use the oceans to produce food if they are to be of great use for satisfying human demands for food. Catching wild fish is not likely to produce the amounts of protein we need. Growing fish in mariculture facilities (marine aquaculture) is an attractive option. Mariculture is particularly intriguing because producing animal protein on land is extremely water intensive; water is a limiting factor for livestock production. However, farming fish and shellfish in the ocean has its own problems.

Current mariculture methods of farming fish and other marine organisms from the ocean are not sustainable. The methods of production rely heavily upon feeding with fish protein caught elsewhere in the ocean and brought in to feed the caged animals being grown. However, easily harvested species are already overexploited. Worse, the original problem for human nutrition is adequate protein. Catching and feeding protein to other animals to grow them for a more desirable form of protein is not an efficient mode of food production. Alternatively, fishes that eat seaweeds and other algae (herbivores) could be grown and fed plant material. This already occurs in Asian freshwater aquaculture, and the brackish-tolerant tilapia needs little animal-derived protein. Most marine shellfish filter materials out of the water and do not need supplemental fishmeal or oils to grow.

Much mariculture is currently a polluting enterprise. Substantial improvements in environmental impacts would be necessary if large-scale mariculture is to provide substantial amounts of food without unacceptable amounts of environmental damage. Much of the diversity of the world's oceans is close to the coasts, the same areas where mariculture is feasible.

People tend to eat large fishes, and those fishes eat smaller fishes. Scientists refer to this as eating high on the food web. Substantial amounts of energy go into growing very large fishes that eat medium-sized fishes, that eat smaller fishes, that eat small shrimps, and that eat microscopic plants. Each step up the food web uses more energy. This is food energy that could be used to feed people more efficiently. Furthermore, organisms high on the food web concentrate pollutants. Mercury is a substantial contaminant of marine fishes worldwide [41].

The lower on the food web humans eat, the more energy is available to feed humanity and the lower the concern about concentration of pollutants. Mariculture is no exception; we will need to start eating seaweed, crustaceans (like small shrimp), and smaller fishes if we want to maximize the amount of food production from the oceans while minimizing toxicants in our food.

How about simply cutting down more forests and creating more cropland and pasture? This is exactly what is going on currently in many developing countries. However, most of the high quality cropland is already under cultivation. Many of the lands left are too steep or have shallow soils that will give poor crop yields. A second problem is that many tropical soils are very poor quality and when farmers cut tropical rain forests for cropland, the fields only provide a few years of crops. Historically some farmers dealt with the quick loss of productivity in tropical soils by cutting down small plots of trees and using the fields for a few years and then abandoning them to the jungle. This slash and burn approach to agriculture requires leaving most of the area not under cultivation and would not provide very high yields. Another problem with creating more cropland in tropical areas is that much of the world's biodiversity is in the remaining tropical rain forests [42]. If we cut these forests for cropland or pasture, we will lose many species forever.

Additional threats to producing crops are also threatening our ability to feed the world. Urbanization is covering valuable cropland. Increased demand for solar power is interfering with cropland in some areas. The emphasis on biofuels has diverted production of crops for food into crops for energy. We divert ever more water from agricultural uses to domestic and industrial uses.

The Green Revolution Rebooted: Is Genetic Modification Our Savior?

Up to about the 1960s, it was obvious to many people that humanity was going to become food limited. The rates of crop production simply were not great enough to supply the rapidly rising demand of a growing human population. The "green revolution" changed all that. Use of fertilizers, mechanization of agriculture, breeding for increased crop yields, and pesticides vastly increased our capacity to produce food. We need to repeat this feat to keep up with continued increases in human population.

One area that has great potential for increasing food supply is decreasing food waste. Approximately 1/3 of food produced is lost or wasted. In developing countries, much food is lost to spoilage and pests. In developed countries much is simply thrown away for various reasons. Increased efficiency of food use will decrease global hunger.

A comprehensive report on increasing crop yields suggested demand for all cereals (maize, rice, wheat, barley, etc.) will increase by 44% between 2008 and 2050. Increases in yield (amount produced per unit area per year) between 1990 and 2010 were 1% each for rice, wheat, and soybeans. This means if yields per unit area continue to increase, we should be able to feed most people without dramatic increases in total crop area [43].

It will be possible to increase application of fertilizers in many areas of developing countries, but soils in the breadbaskets of the world have become saturated with fertilizers. Continued additions in these areas are necessary to keep production high, but application at greater rates will not increase yields. Phosphate supplies are finite, and eventually we will run out of the phosphate we mine for fertilizers [44]. The only thing limiting the production of nitrogen fertilizers is energy as the atmosphere is composed of nitrogen gas that industries can convert to fertilizer with large energy inputs.

Genetic modification (creation of genetically modified organisms) has the promise of increasing crop yields, and the ability to alter genes offers much more rapid modification and in some cases changes that are not even possible, compared to traditional plant breeding. However, the promise of such technologies remains unfulfilled and controversial. A report in 2009 by the Union of Concerned Scientists found that after 20 years of work on genetically modified crops, conventional breeding approaches far outperformed genetic modification [45]. Yields did increase over this period, but most of the yield came from improvements made possible with conventional crop breeding and improvements in other agricultural methods (e.g., harvest, minimizing waste, and decreases in spoilage).

An exciting new development in genetic modification is the possibility that photorespiration in plants can be circumvented. Photorespiration is an inefficiency in the photosynthesis of plants and lowers the amount of energy that can go to crop yields. Scientists have genetically modified pathways in tobacco plants to increase yields by a potential of 40% [46]. Questions remain, such as whether this can be effectively transmitted to food-producing crops and what the ecological consequences are if the same modification moves into harmful weeds, but the promise of increased yields is exciting.

Another avenue to increased yields is protection from losses to agricultural pests. We are making progress on several fronts against such loses. One issue is that pesticides must continue to be developed as agricultural pests evolve the ability to resist the effects of the toxins we apply to stop crop losses. That is, we are currently in an arms race with the pests that threaten our crops (as was discussed in regard to diseases of humans in the last chapter). For the most part, we have been treading water with respect to stopping pest damage. This is one area where genetic modification has seen widespread application.

The first example of progress is putting genes for herbicide resistance into food crops (corn and soybeans). This allows us to plant crops and then apply an herbicide to control weeds that has minimal effects on the crop with the genes for resistance. The problem with this approach is that many weeds have also evolved resistance to the herbicides. Again, we are in an arms race with the weeds, and it will take clever research just to stay even in this race.

The second example is a gene that agro companies have introduced into corn that allows them to resist caterpillars that cause damage. Scientists moved the BT genes from bacteria into the plants (about ten different genes have been used). This allows synthesis of a protein by the plant that eats holes in the guts of the insect larvae that ingest the plant. This BT-producing corn has been adopted by farmers worldwide (except in those countries that have banned genetically modified crops). Of course, the evolutionary arms race continues; the insects have evolved resistance to the BT protein, rendering the genetic modification ineffective in many areas.

Crop yields over the last 30 years have increased. However, they are increasing linearly, and the demand for food is increasing much more rapidly. This will lead to a food gap developing over the first half of this century. While technology has made some impressive increases in crop yields, those increases are not matching recent increased demand. Global analyses also suggest that yields over the last 10–15 years have leveled out in many areas and in most areas of the world have not increased or are even collapsing [47]. We may be able to feed all the people for the next 20 years by maximizing yields with careful use of water [48].

Water: The Most Precious Commodity

Without water, there is no agriculture. Humans are currently appropriating about half of all the readily available freshwater on Earth for our uses. Precipitation that falls on sparsely populated areas, and that which falls faster than people can use it (e.g., in floods), is not available. Ocean water is not a useful source of freshwater for most people (although some desalinization does occur). As such, there is a clear limit on the amount of water available to humanity, and we are using a substantial portion of that water.

According to the United Nations Water Initiative [49], currently 70% of all water use is for agriculture. Whereas each person needs about a half to a gallon of water to drink per day, it takes a thousand times that much water to produce enough food to feed a person for a day. This number is far greater if grain-fed livestock are a significant part of the diet.

About 1/3 of crops are irrigated from groundwater, and the proportion is increasing. The rates at which groundwater is being used globally far exceed sustainable rates [50]. The amount of groundwater available in an area depends on how quickly water recharges the aquifer (i.e., flows down underground from rain on the surface) and how quickly it is lost depends upon withdrawal rates. In most areas where people are using groundwater for irrigation, the withdrawal rate is greater than recharge rates, and levels of groundwater are getting deeper. As a farmer uses more groundwater, the deeper the water level drops below the surface of the land, and the more it costs to pump it up. In the worst case, pumping the aquifer completely depletes it. Aquifer depletion is happening around the world.

Additional problems come with overexploitation of groundwater for agricultural uses. In large areas of India and Bangladesh, there is massive contamination of groundwater by arsenic causing large-scale poisoning of local populations. Many species of freshwater organisms are being lost as streams and springs stop flowing when the groundwater below them is taken away and they are sucked dry. Moving completely to groundwater exploitation makes people less likely to be able to afford water for drinking as the cost of pumping from deep underground makes water more expensive.

Water is likely to become one of the most limiting factors for how much food we can grow. Breeding for more efficient use of water by plants, using species with lower water requirements, and more efficient methods of application will be necessary. Less reliance on meat in the average diet would also improve water use problems. Unfortunately, physics dictates that some potential solutions to water shortages are almost impossible. Moving water any distance requires huge amounts of energy. Desalinating water from the ocean also requires a substantial energy input. Essentially, it is easier to grow food where there is ample water and ship the food, rather than ship the water to grow the food.

Eroding Our Ability to Produce Food

The dust storms of the 1930s caused some of the worst times of poverty, hunger, and despair in the United States [51]. Dust storms rolled across the Great Plains reaching the East Coast of the United States. Plowing the grasslands and the subsequent drought led to mass erosion and loss of topsoil. The major environmental disaster led to adoption of some soil conservation measures, but we still have not stemmed the loss of valuable topsoil [52].

Crops do not only need water, but they also need fertile soil. Soil erosion is one of the least appreciated threats to our ability to increase food production to feed the planet's people. Over time, soil builds fertility from the remains of the plants and microbes in it. The richest deep soils best suited for agriculture took thousands of years to develop. Erosion from water and wind causes soil to be lost. When we plow land and de-vegetate, the surface loss rates increase. Overgrazing leads to increased erosion rates and interferes with the ability to grow livestock. In severe cases, desertification makes it so that few plants can grow at all.

Currently, human activities accelerate erosion rates about 40 times over natural erosion rates, leading to net loss of productivity on farmland. Thus, we are causing our productive lands to become less productive when we need them the most. As we feed more people and develop marginal lands, the rates of erosion will only increase [53]. Overgrazing, wood harvest, and agriculture intensify dust storms originating in several areas on Earth, including China and the Sahara Desert. These dust storms deposit much of the fertile soil into the ocean where it sinks and ends up buried on the bottom of the sea.

Practices to stem erosion of topsoil and maintain crop productivity will be required to feed humanity. Most of these measures are local to regional in nature. They require a long-term view of resource management.

Political Obstacles to Feeding People

Feeding the 9–10 billion people expected to be alive on Earth in 2050 will be very difficult. Global population is actually increasing faster than predicted. Experts think that sustainable agriculture will be able to feed everyone, but it will require global as well as interdisciplinary cooperation.

In his book, *The Plundered Planet* [54], Dr. Paul Collier talks of why the "bottom billion" of the world's population has missed out on global prosperity, including food security. Part of his point is related to how expensive food disproportionally hurts the bottom billion. He cites impediments to production (e.g., bans on genetically modified food), misguided policies in developed countries (subsidizing use of food crops for fuel production), science (undeveloped cropland in developing countries), and the lure of quick gains by those in governments of poor nations (plunder of natural resources for personal gain) as reasons for our inability to feed all people.

Global cooperation in distribution of food, technology increases (plant breeding, pest control), methodological improvements in sustainable crop production (erosion control, improvement in water efficiency), and changes in the way we eat (e.g., less reliance on grain-fed meat) all will be required to feed people into the future [55]. Social stability likely depends on there being some minimal portion of people who are not hungry. Riots occur when food is not available or becomes more expensive. As many as one in seven people on Earth are malnourished at present; apparently this is not enough to cause destabilization of global political systems.

Chapter 6
Nuclear Weapons

Sennacherib's Army Is Destroyed Paul Gustave Louis Christophe Doré, Doré's English Bible, 1966

© Springer Nature Switzerland AG 2019
W. Dodds, *The World's Worst Problems*,
https://doi.org/10.1007/978-3-030-30410-2_6

Hiroshima

At 8:15 on a Monday, August 6, 1945, a US bomber dropped a nuclear bomb on the Japanese city of Hiroshima. The bomb exploded 2000 feet above the city and immediately destroyed the city for a half mile around the point of detonation. The fires triggered by the explosion burned a ring another mile around the detonation point, a circle of destruction of 4.4 miles. About a third of Hiroshima's population, 70,000–80,000 people, were killed immediately. Burns and radiation killed at least another 100,000 people by the end of the year, and additional deaths were attributed to radiation exposure ensued in the following decades. The most terrible weapon the world had ever seen now clearly pointed the way to dominate with ultimate military power.

Following World War II, the USSR (now Russia) and the United States rapidly developed nuclear arsenals. The first bomb detonated at Hiroshima was a fission bomb and yielded the equivalent of about 15 thousand tons of TNT. Newer nuclear weapons release substantially more energy by harnessing the power of fusion. The largest bomb ever detonated, the Tsar Bomba, had a yield of 50 megatons or about 3000 times more powerful than the bomb dropped on Hiroshima. We now have huge numbers of nuclear weapons around the Earth.

There are two major worries. The first is full-out nuclear exchange that destroys many major cities and makes the rest of Earth an uninhabitable radioactive soup. The second is a limited nuclear exchange that will put enough dust into the atmosphere to cause drastic cooling for a shorter period. I will consider both these issues in this chapter.

What Are Nuclear Bombs?

The first nuclear bombs discovered in the 1940s are referred to as fission bombs. Each atom of any element except hydrogen has a nucleus composed of protons and neutrons. Radioactive compounds generate energy when their nucleus splits. Some radioactive elements release large amounts of energy because their nuclei are so unstable and they break down relatively rapidly. Two such compounds are uranium and plutonium. If you put enough of either of these elements together in one place, the energy generated from breaking down one nucleus can cause one or more other nuclei to break down, releasing even more heat. This is a chain reaction.

In nuclear reactors, this reaction is controlled, and a nuclear plant uses the resulting heat to generate electricity. In nuclear bombs, the reaction is uncontrolled; it generates so much heat so quickly, and it causes an explosion. The trick in making a powerful nuclear bomb is to make all the radioactive materials react before they blow themselves apart. Weapon makers construct the bomb by surrounding the radioactive material with conventional explosives that smash the radioactive material together and start the chain reaction.

By 1952, scientists had created the fusion bomb. In this bomb, two forms of the simplest element, hydrogen, fuse to make the element helium. Thus, the bombs are hydrogen or H-bombs. When fusion reaction occurs, there is a tremendous release of energy and release of very energetic neutrons (one of the particles that make up the nucleus).

In a fusion bomb, the older type of fission bomb reaction compresses the two forms of hydrogen. The two forms fuse, and the fusion reaction creates neutrons that feed energy to other nearby fusion reactions. The fusion reaction also accelerates the fission reaction and can interact with other elements in the bomb to cause fission reactions. Thus, H-bombs really are a hybrid between fusion and fission reactions, but the huge amount of energy released by the fusion means these bombs can be much more powerful.

There are several other types of nuclear bombs related to these two types. One is a neutron bomb, a fusion bomb that weapon designers have optimized to produce little radioactive fallout. It does produce an intense burst of short-term radiation by releasing many neutrons. This type of bomb could kill many people without destroying nearly as many buildings and other structures. Presumably, such a weapon would allow an aggressor to take over its target for its own uses.

A dirty bomb does not rely on fission, but uses a conventional explosion to release deadly and poisonous radiation. Plutonium, commonly used in nuclear bombs, is very toxic as it concentrates in bone marrow where it can cause damage to production of blood cells. Similarly, uranium used in bombs and nuclear reactors is as toxic as a conventional poison in addition to causing radiation toxicity.

Global and Regional Nuclear War

In the worst-case scenario, 10,000 nuclear weapons are detonated. It may seem unlikely, but with as many armed nuclear warheads as are available, we are perched minutes to hours away from such a scenario at all times. In all-out war, Russia and the United States would target all cities over 1 million people in each country and many in other countries, killing hundreds of millions of people immediately. These people would be the lucky ones. The remaining people on Earth would be contaminated with radiation as well as suffering the effects of nuclear winter.

According to the experts [56, 57], there are about 13,890 nuclear weapons in arsenals around the globe. This is down from a peak of 70,300 in 1986. These numbers are somewhat vague, as most countries do not want to publicize the number of weapons. Even the United States stopped making this information public under the Trump administration. A number of countries around the world have substantial numbers of weapons (Fig. 6.1).

Currently the human death and health effects of nuclear weapons are minor, as well as the environmental effects. We have learned to protect people from harm that are working with highly radioactive materials. This is not to say that the purification and construction of nuclear weapons are without negative health effects or

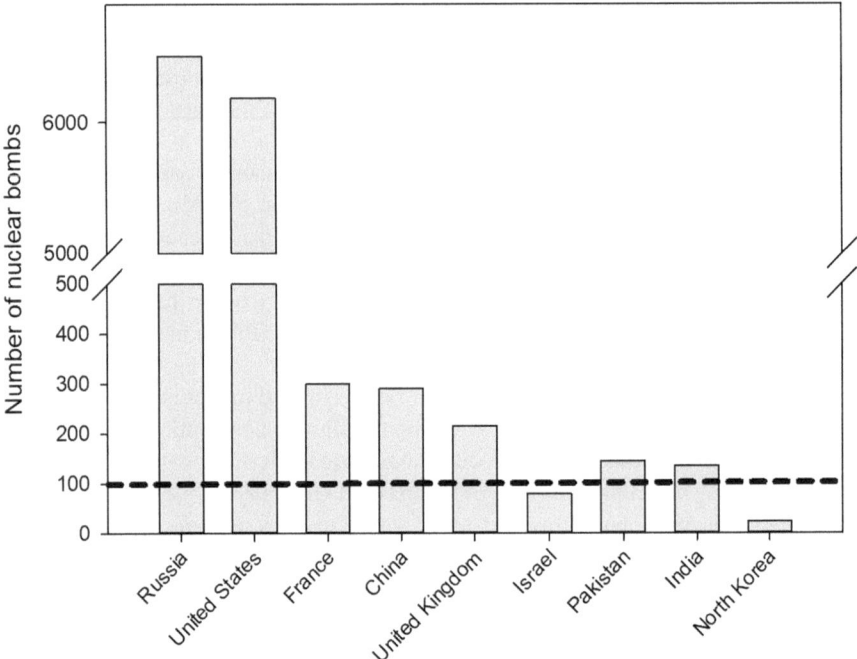

Fig. 6.1 Total stocks of nuclear weapons by country in 2019. Note break in scale from 100s to the 1000s that the United States and Russia have. The line is at 100 weapons, where global atmospheric effects are expected if that many weapons are detonated. (Data from the Federation of American Scientists downloaded 7/2019 https://fas.org/issues/nuclear-weapons/status-world-nuclear-forces/)

environmental problems, but rather that relative to other global sources of mortality, suffering and environmental damage harm *relatively* few people, with damage limited to small areas.

Radiation

Both fission (atom) and fusion (hydrogen) bombs rely on fission reactions to create explosions. In hydrogen bombs, the fission explosion is controlled and leads to the fusion reaction, but both types of bombs create huge amounts of tiny radioactive fragments that atmospheric circulation spreads around the world.

We know that the radioactive particles (isotopes) spread around the world because countries used to test nuclear weapons above ground. These explosions created huge mushroom clouds that injected radioactive materials high into the atmosphere that then circulated around Earth. We can still detect the radioactivity from these explosions, and it entered the bodies of all people on Earth.

The nuclear isotopes created by these explosions break down, and the amount of time required for them to break down is their half-life. The half-life of a radioactive isotope is the time it takes for half the material to break down. Half-lives of isotopes from thermonuclear explosions range from days to thousands of years. Cesium, strontium, and iodine isotopes that are produced and spread by nuclear explosions are of the greatest concern because we can readily take them into our bodies where they can cause rapid poisoning in higher doses and greatly increased probability of cancer at low doses.

Lethal doses of radioactivity from a one megaton nuclear device would extend 20–40 miles from the point of explosion, with amount of time of exposure to the fallout required for a lethal dose increasing further from the explosion. Still, if most major cities in a densely populated area, such as the East Coast of the United States, were the targets of a nuclear attack, most people between those cities would receive lethal doses of radiation even if the explosions did not kill them immediately. Full-out nuclear war would substantially increase cancer rates worldwide as the atmosphere disperses isotopes. Scientists have understood this effect for over a half century, and this eventually led to international agreements to cease atmospheric testing as the negative global impacts of radiation releases into the atmosphere were understood.

Zapping the Computers

One effect of nuclear explosion is the release of an electromagnetic pulse that damages most normal electronics. This complex effect depends on, among other things, the size of the nuclear explosion and the altitude. Due to interactions with the upper atmosphere, a large nuclear weapon detonated 250 miles above the center of North America would damage most electronic systems. This would include, but not be limited to, aircraft and satellites.

Even partial damage to electronic equipment could disrupt basic services such as food distribution, power distribution, water, monetary exchange, and communication. Most parts of most countries are dependent on computers that would be inoperable after exposure to the electromagnetic pulse from a nuclear explosion. Disruption of these support services would very rapidly lead to suffering unless we could repair the infrastructure quickly. For example, perhaps only a week or so worth of food is stored in most metropolitan areas. If food, heat, power, and clean water (no power to pump and purify water) are gone, people would quickly start starving, and communicable diseases would increase rapidly.

Ozone Depletion and Sunburn of Earth

Nuclear explosions have other potential negative effects on the Earth's capacity to support humanity. For one, they could lead to partial to complete destruction of the protective layer of ozone in the upper atmosphere that filters out most harmful ultraviolet rays before they reach the surface of the Earth. A respected study [58] calculated the effects of 100 small nuclear detonations occurring in the subtropics. This number of explosions is within the range of possibility if a nuclear exchange occurred between India and Pakistan. Of course, a larger scale nuclear exchange would have more extreme effects. One hundred small warheads would represent detonation of less than 1% of global nuclear stocks. The predictions for catastrophic effects on the world's crops still hold as climate models continue to improve.

Nuclear explosions light cataclysmic fires in a large ring around the blast. These fires and the heat of the explosion cause a huge column of smoke and superheated air that moves high into the atmosphere. We are all familiar with the mushroom cloud produced by nuclear explosions; the mushroom cloud proves the capacity of such explosions to inject material high into the Earth's atmosphere.

The fires from the explosions would inject large amounts of smoke consisting of very fine soot particles into the upper atmosphere. The soot would absorb the sun's rays and heat the upper atmosphere (while cooling the lower atmosphere, which I will discuss next). The heating allows the reactions that break down the protective ozone layer to occur at greater rates than normal.

Thus, a nuclear war with about 100 warheads would lead to a 20% increase in UV in the tropics, about a 50% increase in the middle latitudes, and around 75% near the poles. UV light is damaging to people (increasing cancer rates) but also would be very damaging to plants, including food crops. The damage would go on for at least 5 years. I will discuss ozone depletion more completely in its own section.

Observations following the 1991 volcanic eruptions of Mt. Pinatubo (the Philippines) and Mt. Hudson (Chile) have confirmed the effect of large explosions on climate. These volcanoes led to a 15–20% ozone loss at high latitudes. They also led to global cooling, discussed below. Nuclear explosions should lead to longer-term effects than volcanic explosions because the soot particles produced by the nuclear explosion are far finer than the ash produced by volcanoes. The finer particles take longer to settle back out of the upper atmosphere.

Nuclear Winter and then Nuclear Summer: Out of the Fridge into the Fire

Ozone depletion is not the only problem that would occur with respect to the Earth's capacity to support humanity. Scientists also predict a "nuclear winter" followed by a very hot period, a "nuclear summer." Both of these effects would take a heavy toll on the natural systems that support humanity and have global effects, reaching areas far from the area where the nuclear blasts actually occurred.

In 1815, Mt. Tambora erupted in present-day Indonesia; it was probably the largest volcanic explosion in the last 10,000 years. This eruption was 10 times stronger than the Mt. Pinatubo explosion and around 100 times stronger than the explosion that blew the top off Mt. St. Helens in the United States. The explosion resulted in global cooling, with an extremely cold spring and summer in northern latitudes.

Snowfalls and frost occurred in the summer in high latitude areas, and Northeast United States and Europe saw failures of most crops. Hunger was widespread.

Atmospheric scientists knew of these effects before development of extensive nuclear arsenals and predicted that nuclear war could lead to similar conditions. More recently, scientists have used global circulation models (the models used to predict global warming from greenhouse gas emissions) to calculate the effects of nuclear war [59]. The models suggest that all-out nuclear war would lead to an average global cooling of about 14 °F for several years and even a decade later the cooling would be about half that. This extreme cooling (more extreme than the last global ice age) would kill many plants around the world not adapted to the colder temperatures. Iowa and the Ukraine would freeze for an entire year. Agriculture would become extremely difficult in the mid-latitudes that are currently the bread-basket of the world. Three years after the explosions, cooling would still shorten growing season in most of the United States and Europe by 3 months. Even a limited nuclear exchange would cause global cooling and some crop failure. Global starvation would result in parts of the world where people were able to survive the initial effects of the nuclear explosions and subsequent radiation poisoning.

The drastic decreases in temperature would probably kill many plants on Earth. Tropical plants that are not able to withstand freezing would certainly die. Temperate plants would either be completely frozen or experience freezing temperatures at times of year they are not adapted to. Ultimately, the climate would warm as the particulate material settled out of the atmosphere, and the dead vegetation would start to decompose. The decomposition would result in carbon dioxide production. Usually carbon dioxide release by decomposition is balanced by uptake during the process of photosynthesis. However, if many of the plants are killed by cold, or injured by increases in ultraviolet from ozone depletion, there is limited photosynthesis, and there would be a strong increase in carbon dioxide in the atmosphere. Additionally, the huge fires from the nuclear explosions would lead to the primary product of combustion, carbon dioxide. The fires would immediately increase atmospheric carbon dioxide, and this would take decades to leave the atmosphere.

In a nutshell, increases in atmospheric carbon dioxide would cause the entire atmosphere of Earth to warm and lead to increased surface temperatures (I will discuss this in more detail in the next chapter). Thus, once the soot particles settled out of the upper atmosphere, and nuclear winter eased, the globe would warm drastically. This nuclear summer would follow the nuclear winter. One can imagine ice and snow building up for a decade and then a very rapid warming leading to catastrophic flooding.

Post-fusion World?

Given the drastic effects of full-scale nuclear war, the resulting world would be very different from the Earth we inhabit today. Only the hardiest plants and insects would be able to survive the radioactive radiation, the increased exposure to ultraviolet rays from the sun, and the drastic swings in temperatures. Essentially all large animals and most large plants would disappear. If some people were able to hole up for several decades until the Earth once again became habitable, there would be little capacity for food production, and the Earth would be a very desolate place. It is not likely that humanity would survive full-scale nuclear war.

Even a modest exchange of nuclear weapons would lead to huge numbers of dead and suffering with effects rippling out around the world. While scientists have known for many years the apocalypse that would befall humanity with large-scale nuclear exchange, the negative effects of a modest exchange are worse than we thought previously [60]. There would be global suffering, mostly through increased ultraviolet rays and global cooling. The number of people that are already experiencing hunger in the world is large, but a serious blow to the world's food supply would make the problem much worse and lead to even more suffering.

A difficult aspect to predict, because human behavior is difficult for us to predict, is how well societies would survive the destabilizing effects of severe shortages of food and a massive depression of the world's economy. The global economic downturn in 2009 increased the amount of poverty and suffering in the poorest countries as well as pushing more people in developed countries into poverty. We can only speculate as to the total amount of suffering related to societal collapse after limited nuclear exchange, but it certainly would exceed anything seen in recent human history.

Nuclear Terrorists

A more recent concern is the possibility that a nuclear weapon or weapons will fall into the hands of terrorists. In this case, the most likely scenario is that terrorists would detonate one or a few weapons in a large city. The death toll would be considerable, with millions of people either killed immediately or dying from burns or radiation exposure following the blast.

The considerable technology required to produce a functioning nuclear weapon makes it unlikely for a terrorist group to produce a nuclear weapon on their own. Naturally occurring forms of uranium are used to create a nuclear reactor, and then the reactor can give rise to the form of plutonium that can be used in bombs. The nuclear reactor must be operated in a fashion different from the way it is used for power production. In addition, concentrating the form of plutonium that makes a nuclear bomb requires sophisticated mass-scale chemical procedures. All these technical requirements make it possible to determine when a country is producing weapons-grade plutonium, and adequate and thoughtful international intelligence should be able to detect such an effort if well-funded terrorists attempted it. It is far more likely that a terrorist organization would steal a bomb that was already made. This was a real fear following the fall of the Soviet Union.

Creating a "dirty bomb" is more within the realm of possibility for an individual terrorist group. In this case, the terrorists would use a conventional explosive to spread radioactive material through a city. The death and suffering would be horrific and almost completely related to radiation effects. Some would die immediately from radiation poisoning, but many would die of radiation-related diseases in years to follow. Such bombs would have a large economic effect in addition to the immediate death and suffering they would inflict. Think about the economic influence of the September 11 terrorist attacks on the World Trade Center. Even a small nuclear weapon would inflict far more damage than the September 11 attacks.

Stopping this kind of attack requires accounting for and protection of large amounts of nuclear materials. Such accounting and protection become more difficult when there are numerous nuclear reactors because the fuels from these reactors, while not suitable to produce an atomic explosion, are poisonous and radioactive enough to make a very dangerous "dirty bomb."

Perhaps the largest effect a nuclear terrorist attack could have would occur if a bomb was detonated several hundred miles above a country, resulting in damage to electronic equipment over a radius of a thousand miles. Any aircraft in the air in that region at the time could crash, as they would lose their electronic systems. The resulting effects linked to decreased food production and distribution and economic stress would influence the entire world.

Delivery of such an explosion would require a technologically advanced intercontinental missile to transport and explode a high-yield nuclear warhead at a high altitude. It is unlikely that a terrorist group could produce such a device on their own. It is not out of the realm of possibility that a group with a very technologically advanced member could gain access to an existing nuclear missile facility in one of the countries that has such capabilities. There are nine countries (China, France, India, Israel, North Korea, Pakistan, Russia, the United Kingdom, and the United States) that currently have both intercontinental ballistic missiles and nuclear warheads. Some of these are not extremely stable regimes and may be vulnerable to terrorist takeover of nuclear missile capability.

While these nuclear terrorist scenarios are scary and terrible, they do not lead to the equivalent or greater death and suffering as a number of the other of the world's worst problems that I discuss in this book. Nuclear terrorism could be more devastating than the Tsunami in Southeast Asia in 2004 that killed 230,000 people and led to widespread economic damages. Unlike nuclear terrorism, a modest nuclear exchange (a few hundred) or full-scale nuclear exchange (1000s) of detonations certainly can be considered one of the world's worst problems given the likelihood of global death and suffering.

Chapter 7
Global Environment in the Anthropocene

Hebo (河伯), the god of the Yellow River, riding on dragons and turtles to go back to his palace, Xiao Yuncong, 1645

© Springer Nature Switzerland AG 2019
W. Dodds, *The World's Worst Problems*,
https://doi.org/10.1007/978-3-030-30410-2_7

If a scientist studied our planet 1 million years from now, would they see signs of the human activities occurring now on Earth? The answer is yes, and most of what they would see is due to our activities over the last 100 years. We are altering numerous aspects of the global environment so radically and widely that scientists are referring to current times as a new geological era, the Anthropocene [61].

What traces of our current civilization would remain far down the road? How can we be influencing an entire planet? Our large concrete and steel structures will eventually get buried and preserved (fossilized) for future scientists to uncover. Increasing atmospheric CO_2 is decreasing the ocean's pH (increasing acidity). This leads to different rates of chemical deposition into the materials accumulated on the bottom of the oceans and alterations in the species composition of the oceans.

Some argue that humanity and our actions are insignificant relative to how large our world is; this viewpoint is becoming impossible to defend in the face of facts. Vaclav Smil documents how the human population has increased by 30-fold, while the total amount of plant material on Earth has decreased by about half in the last 2000 years. In the last 100 years, total mass of wild mammals has decreased by half (by total global weight or biomass), and cattle alone have increased fourfold [62, 63]. Domestic livestock biomass now exceeds wild mammal biomass by over 20 times [64]. People have converted tremendous areas of land for crop production.

We are causing dramatic global shifts in the way that our ecosystems work and provide food for humanity. We are causing the sixth mass extinction of global diversity; the loss of over half of all species on Earth in a geological timeframe that is just a blink of the eye will be clear in the fossil record. Furthermore, we have shifted marine fisheries such that very few large fish remain in the open ocean. Marine fossils from our time discovered far in the future will reflect this decrease in diversity.

Humans are now the primary mechanism of movement of soil on Earth. We plow the land and alter vegetation, and erosion rates far exceed those seen for many thousands of years. Relative to all geological processes such as weathering, wind, floods, and tectonic and volcanic activities, we are now the dominant force shaping the way the surface of the terrestrial Earth looks. The eroding materials created by our actions enter the oceans, creating a clear signal that the ocean sediments will preserve for millennia.

Atmospheric testing of nuclear weapons from 1945 to 1963 spread radioisotopes around the Earth, and these radioactive materials will form a distinct layer in sediments preserved into the future. From the vantage point of the far future, people are having an unprecedented influence on the global environment, an environment that has supported life for billions of years. Even areas that are "protected" are under human pressure from atmospheric pollution, human incursion, and other human influences. This leaves only the most remote northern and southern regions of the globe as relatively pristine [65].

Is it good or bad that we are changing the global environment? One way to quantify our influence is through our ecological footprint. The footprint is how much of the Earth's resources (expressed as a geographic area) we use to support our current population. If we exceed the area of the Earth that can support each activity, then we are using up part of the environment that could support people in the future. The idea that no person should suffer disproportionately now, or in the future, suggests that a footprint that exceeds the capacity for support is taking away from future generations.

A business serves as an analogy for how we can use ecological capital and the short-term versus long-term view. On the one hand, you can build a business for long-term health; you build the capital of the business to weather bad economic times and perhaps to pass it on to your children. On the other hand, you deplete the capital holdings of the business and operate riskily; increasing the possibility it could collapse catastrophically. Venture capitalists use the extreme of this strategy; they allow a company to collapse, sell it off, and use the money elsewhere for more profit. There is no other Earth if we deplete our environmental capital to the point of collapse. This is a cold fact. So the question becomes to what degree are we actually depleting our environmental capital and is collapse possible or even probable?

Our Footprint

The global footprint is how much land is required to support humanity [66]. When people were hunter-gatherers or agriculturalists before mechanized agriculture and transport, the area of land required to support each person and the amount of water needed were clear. Each person lived near the land that supported them and the consequences of overexploiting the land were obvious and immediate. It becomes more difficult to calculate individual footprint as societal interactions become more complex, and material and labor transport become more rapid. However, accounting for global use of resources (independent of how they are distributed) is possible.

Currently, we use half of the freshwater available on Earth as I have discussed in Chap. 5 with respect to how much water is needed to feed people. The water cycle continuously supplies clean water from rain and snow. However, not all of it is available for human use. Some falls far from places where people need it (such as in the Arctic). It is not practical to move water long distances because it is very heavy and expensive to move. Much water is lost in floods; we simply do not have the capacity in reservoirs to catch all this water. Of the remaining water, we appropriate half of what the hydrologic cycle supplies to us each year. Some water evaporates, particularly with agricultural uses, and we cannot access it until it falls as precipitation. If we used it for domestic and industrial needs, it can be contaminated and costly to clean. As far as humanity is concerned, there is a definite limit to how much water is available for our use.

Demand for water is increasing very sharply for several reasons. One is that the population on Earth is growing rapidly, and we need to water crops to feed everybody. Another is that the standard of living is increasing around the world [67]. It takes water to create all the goods, special foods, and luxuries people in developed countries have come to expect. Every pound of grain-fed beef takes 11,000 gallons of water to produce. In contrast, it takes 106 gallons to produce a pound of wheat. Thus, as we eat richer diets, we increase the demand for water. It takes about 40,000 gallons of water to produce a car. Industry also has a large demand for water.

Economic growth is required to lift people out of poverty. The amount of economic growth is directly tied to energy use [68]. Energy use can have strongly negative environmental effects (e.g., the greenhouse effect). Thus, increased demand for energy to fuel the economy creates difficulties for protecting the planet.

Feeding people also takes land out of its natural state and converts it to cropland. We have used much of the land available for agriculture, though estimates vary [69]. The rest is too rocky, cold, or inaccessible to cultivate profitably. While there still are areas that can be cultivated, there are clear limits to the amount to arable land on Earth, and we are approaching those limits. Some of the remaining areas include tropical rain forest that humanity is cutting at alarming rates to create crop and pasturelands, so there is a trade-off between protecting biodiversity and increasing the amounts of cropland and pasture.

One of the largest consequences of humanity's footprint is what we are taking away from other species. With habitat destruction and overharvesting, we are using up nature at unsustainable rates. The rates of human-caused extinction exceed rates of evolution of new species by 100–1000 times [70]. We are causing extinctions at such a great rate; it will take millions of years for an equal number of species to evolve to replace the ones that have been lost. Thus, loss of species caused by the actions of people is for all practical purposes not reversible; the humans on this planet are causing a permanent global effect on the rest of the organisms on Earth.

The footprint idea leads to the idea of safe operating space for humanity on a finite planet. This view considers all the resources we use, their impacts on the environment, and the upper boundary of use rates that Earth can sustain. The idea of planetary boundaries is essential to understanding the finite ability of the Earth to support humans into the future [71].

Economic Value of Maintaining the Global Environment

In the 1990s, New York City was up against major projected costs to meet drinking water standards. In response to these projected costs, the city took an unusual path. The city purchased and protected land in the 2000-square mile watershed that supplies the city with 90% of its water. Buying land, repairing septic tanks, helping upstream communities with sewage treatment plants, and developing other measures to protect the water cost $1.5 billion dollars. In contrast, planners estimated that building water treatment plants to purify all the unprotected water leaving the watershed would cost $6 billion up front and $250 million per year in operating costs. Thus, protecting the water had specific economic benefits for the city and its inhabitants [72]. Other cities might learn from this example!

This case is an exception, and traditional economists took for granted many of the benefits that the global environment produces for humanity on a daily basis. They termed these externalities, which meant they were external to the ability of economists to include them in the economy. However, this position was not so hard and fast even for old-school economists. People pay for fish taken from water and from freshwater. Many of us pay for the domestic water we use directly from the tap, and people buy bottled water to drink. In dry parts of developed countries, water fetches a large price. However, in the face of traditional economics ignoring much of the value of the natural environment in sustaining humanity, a group of ecological economists (notably Robert Costanza and his colleagues) started the process of assigning economic values to the things that the environment does for us [73].

Some of the values are straightforward, but it takes a bit more ingenuity to establish others. Take, for example, the value of clean water. If we want to use polluted water, there are technologies to clean it, but it costs money (energy, chemicals, and equipment) to purify. Thus, a person who contaminates water is decreasing its value. Conversely, there are processes that naturally clean water (e.g., passing through a wetland) that an intact environment provides.

Hurricane Sandy struck the East Coast of the United States in 2012. Many municipalities were without clean drinking water because they did not have the electricity to run purification plants. New York had protected much of its water by preserving the forested watersheds that produced its water and designed much of its system to be gravity fed. Thus, most residents had clean drinking water. The benefits to buying land and protecting the watershed above the city were even more than the initial calculations suggested.

Water is my specialty area, so I will continue with examples of the value of water. We can pollute water with nutrient runoff from cropland, feedlots, and cities (the nitrogen and phosphorus fertilizers that are in runoff from fields and human and animal waste). When the water runs into lakes and oceans, it creates conditions that favor growths of high concentrations of algae (single-celled suspended "plants"). The water turns green (generally people don't like this) and smells bad. Waterfront property values drop. Furthermore, if we use the lake for drinking water and the smell and taste of the water is bad, removing the odor is expensive. Many city water systems do not remove all the taste and odor, so people spend money on bottled water instead of using water from the tap. Communities lose tourist dollars if the lake is a recreational destination and water pollution drives away business. In more severe cases, the pollution fouls the water so badly that the fish die. Finally, the chance of toxic algae increases with heavy nutrient contamination, making the water unsafe for drinking or contact. Such pollution in lakes has led to deaths of domestic livestock and pets and had negative human health implications for those that rely on the water [74]. Modest expenditures to curb nutrient runoff can prevent substantially greater economic costs associated with property values, recreation, and supply of safe drinking water.

I provide logging as another example. There is immediate benefit from logging from the value of the lumber. However, the basic functions of the forest are changed after logging in addition to the time it will take the trees to regrow to the size that the area can be profitably logged again. For example, forests tend to hold in water and increase its purity. This function of forests allows for lower probability of flood and drought downstream as well as providing pure water. This is why the city of New York was willing to pay to preserve forests upstream from the city's water supply; it made simple economic and ecological sense. Forests also serve as a refuge for wildlife, photosynthesize and remove carbon dioxide from the air, and provide an area for people to recreate.

As you can imagine, every ecosystem on Earth has some direct and some less direct benefits to humanity. Simultaneously, human actions are influencing every ecosystem on Earth. We know now that we can quantify what some economists formerly considered externalities, and the services the environment supplies us can become part of the equation of the economy.

An additional issue in economics is the idea of discounting. With this approach, a person should take advantage of economic opportunity immediately if they will lose more money in the end by not exploiting the advantage. Take, for example, a forest of moderate-sized trees. You could cut them immediately and make some money. You could wait and cut them in 20 years and make more. The strictly economic decision would say if you can make more interest on your money if you cut the trees now and put the money in the bank, you should cut the trees. However, this approach can ignore the idea that future generations might value the forest for their own reason, but if you chop it down, they will not have the possibility of enjoying a mature forest for many years to come. Stated differently, cost/benefit analysis into the future is very sensitive to the degree to which we are willing to assume costs today to benefit people in the future [75].

We cannot restore some resources provided by ecosystems quickly once they are used. For example, as we increase the carbon dioxide in the atmosphere by burning fossil fuels, it dissolves in the oceans and causes them to become more acidic. The acidic water does not allow corals to reproduce and grow. If we cause extinction of coral species, they will never come back. The ocean may take centuries or more to recover, but species extinction is forever. Species diversity is tied to ecosystem services that support humanity, so extinction could have long-term consequences in addition to the loss of a species [76]. Even causing formerly abundant species to become rare could have negative impacts on ecosystem functions that support humanity [77].

Natural ecosystems also can be subject to complete abrupt collapse. We can go back to the example of lake and nutrient pollution. As more algae grow in a lake, they sink to the bottom waters. This increases the probability that all the oxygen gas dissolved in the water will disappear. Once this happens, the sediments no longer hold nutrients, and water currents and diffusion continuously mix them back into the lake. Simply stopping the nutrient pollution running in from outside is no longer effective; the system has flipped to a less desirable state. Likewise, fisheries can produce fish sustainably if anglers harvest them at modest rates. However, once the harvest exceeds a certain rate, the fishes can no longer reproduce rapidly enough to replace the harvested animals, and the fishery collapses almost completely. This abrupt change is in direct contrast to extraction of many nonliving resources.

The staggering finding from Robert Costanza and his colleagues is that if you sum up all the values of ecosystem services, they rival the total value of the world's economy [78]. Thus, humanity's well-being requires protection of the environment that feeds and shelters us; our economy links inextricably to the environment.

Of course, the value of our natural world is not just monetary. People get enjoyment out of the natural world. The natural world is sacred to many people. As stated earlier, I am operating under the assumption that plants and animals have an intrinsic right to exist. None of these factors relate to direct monetary valuation. The point of this section is, however, that ignoring the links between economy, environment, and the potential for death and suffering of people is not wise. So, how big of an impact are we having globally on the environment, and how much death and suffering will this cause?

The Greenhouse and the Deniers

One of the biggest worries with respect to global environment revolves around the relationships among increased release of "greenhouse" gasses, global warming, and changes in global climate. This issue has become a flashpoint between ideology and science. There are many books on this issue, so I will only cover it briefly here.

There are several ways people question the idea that burning fossil fuels is leading to global warming. The first: are we increasing the amount of carbon dioxide and other greenhouse gasses (nitrous oxide and methane) in the atmosphere? The second: can these gasses increase temperature? The third: are there other potential causes for temperature increases? Briefly, yes, yes, and no.

How do we know humans are increasing the amount of greenhouse gasses in the atmosphere? It is a simple matter to measure how much carbon dioxide, nitrous oxide, and methane are in the atmosphere. The best record of carbon dioxide measured directly comes from the Mauna Loa Observatory in Hawaii. Between 1958 and 2018, the concentrations of carbon dioxide in the atmosphere have increased steadily from 315 to 410 parts per million (a 30% increase). However, other methods support the observation that atmospheric gasses are increasing. Scientists drill ice cores in Antarctic and Greenlandic glaciers and analyze tiny gas bubbles trapped in them. Naturally occurring chemical forms of other chemicals are also trapped, and these are studied to date the depth of the ice core. The cores show that carbon dioxide never rose above 300 parts per billion over the last 800,000 years [79]. Scientists have obtained similar results for two other important greenhouse gasses, methane and nitrous oxide. So what has changed to allow for these increases in gasses?

The only explanation for the carbon dioxide increases to date is that deforestation and land disturbance coupled with burning fossil fuels (oil, natural gas, coal) have caused the increases. More direct evidence comes from the chemical signature of carbon dioxide from fossil fuels [80]. This chemical signature relates to the fact that elements such as carbon are composed of slightly different versions of the same element (called isotopes). The long-term isotope ratios in the atmosphere from the ice cores are very different from those created by burning of fossil fuels. The atmosphere is currently moving toward the isotopic signature of fossil fuels.

Therefore, we are increasing the concentration of these three gasses in the atmosphere. We have increased carbon dioxide the most in terms of absolute amount, but we have also substantially increased concentrations of methane and nitrous oxide. All three of these gasses are very effective at absorbing heat. Methane absorbs 21 times and nitrous oxide 310 times more heat per molecule than carbon dioxide.

The balance of energy coming in from the Sun and escaping back out to space determines the temperature of the surface of the Earth. If there are more gasses in our atmosphere that absorb heat, they block the escape of heat back into space, just like putting more blankets on your bed in winter. The temperature of the land, the sea, and the atmosphere will increase until enough heat energy escapes into space to balance temperature. Since this is very similar to what happens in a greenhouse, where the Sun's heat is trapped by the glass, and the inside temperature can far exceed the outside temperature, the gasses are referred to as greenhouse gasses.

Svante Arrhenius first proposed the idea that carbon dioxide in the atmosphere controls the temperature balance of the Earth in 1896. However, the idea remained in the background until the 1970s. At that time, the Earth was in a cooling trend, and about 10% of scientists suggested that the Earth would enter another ice age. Still, many climatologists accepted that increases in carbon dioxide would lead to increases in temperature. In the 1980s, the issue bursts onto the political scene, and public debate has been going on since that time.

The scientific community, since the 1980s, has come to stronger and stronger consensus that the Earth is getting warmer and that increased greenhouse gasses are the cause of that warming. Predictions in the 1990s were that temperatures would increase, glaciers and the Arctic ice cap would decrease, the climate would become more variable, and the sea level would rise. Each of these predictions has come true. The statistical certainty that the Earth is warming over the last 40 years is now greater than 99.99999999% [81].

Instrumental temperature measurements allow estimation of global temperature since about the 1850s. Ten of the warmest years on record since 1880 occurred between 2005 and 2017 as determined by direct land and ocean measurements by the Berkeley Earth group that created new methods to compare the many records taken around the globe [82]. Several independent groups have found that the global temperature has increased about 3 °F. Richard Muller and colleagues founded the independent Berkeley Earth Surface Temperature Study, which received some funding from the politically conservative Koch brothers. This research group was initially skeptical of the analyses conducted so far. They reanalyzed all the data and almost exactly matched the estimates previously published by the other groups. In addition, they used methods to extend the record back to 1800 and found an even more pronounced temperature increase [83].

Glaciers worldwide are shrinking. The Arctic ice cap is becoming ever smaller. The very oil companies who have expressed skepticism over global warming and the link to burning fossil fuels are taking advantage and drilling in areas not accessible in the Arctic Ocean previously. The global shipping industry is now able to transport cargo through the Arctic during the summer. Sea level has risen on average over 8 inches over the last 150 years. Scientists predicted all these things, and now they have happened.

There is no other scientifically accepted explanation for global warming. The luminosity (brightness) of the Sun is not increasing over time, the tilt of the Earth's axis cannot explain it, and other weather patterns and cycles are not to blame. While scientists and nonscientists have made several proposals to explain the warming, peer-reviewed scientific literature ultimately has supported none of them.

This issue is one of the most frustrating to me as a scientist observing the political process. It illustrates a case where people often let what they want to believe overrule the scientific consensus. Some states have even legislated "against" greenhouse warming, as if politics can reverse a scientific fact. Of the thousands of papers on the global climate, not one provides good evidence for any other explanation for current trends in global increases in temperature and the resulting global climate change, other than humans increasing greenhouse gasses and this leading to global climate change. Residents of countries that emit more greenhouse gasses per person tend to be most skeptical of the greenhouse effect, and those are convenient views given protecting the environment might interfere with their current lifestyles [84].

What kinds of problems will come about if we do not control climate change? Estimates are that at current carbon dioxide levels in the atmosphere, the Earth's average temperature will rise 5 °C [85]. Five degrees centigrade may not seem so bad, but this is 9 °F! One major problem will be the increased severity and length of heat waves. These events kill many people each year, and this will only get worse. Another problem is the greater temperatures could lead to crop failures, making it more difficult to feed all people. Elevated temperatures also lead to more human interpersonal violence and frequency of intergroup conflict [86]. Sea level has is predicted to rise yet another 8 inches in the next 50 years. Scientists predict more extreme weather events with a more energetic (warmer) atmosphere. As organisms become mismatched to their warmer world, many are expected to be unable to adapt and will go extinct.

Of particular concern are a few potential scenarios where a runaway greenhouse effect occurs. One of these revolves around the permafrost in northern Europe, Asia, and North America melting and releasing the carbon stored frozen in those soils as it built up over thousands of years. The release of gas will cause more warming and more permafrost melting, and this will feedback releasing more gas, causing more warming, and so on.

Another potential for runaway greenhouse is that warming will eventually cause all the oxygen dissolved in the waters at the bottom of the ocean to disappear. This could happen if the warming disrupts ocean currents to the point that the bottom waters mix less rapidly than they do now coupled with human activities increasing the fertilization of the ocean surface waters. The lack of oxygen will kill many organisms and favor processes in the ocean that create even more greenhouse gasses. These are all areas that need more study so we know the probability of them happening and can prepare steps for mitigation.

Yet another potential runaway of concern is the increase in temperature of the oceans, leading to greater rates of methane release. A number of relatively shallow areas of the ocean have large deposits of frozen methane stored in their sediments. Releases of this methane are increasing already in the North Atlantic [87] and probably elsewhere due to warming of the ocean and associated shifts in ocean currents. As shallow marine sediments release more methane, the planet warms more, leading to even more methane release.

Scientists are generally conservative with respect to making predictions into the future that fall outside of their area. While they will predict temperature, sea level, precipitation, and extreme weather increases, they are generally unwilling to predict the social consequences of those changes. Some suggest that climate change will lead to global societal tipping points, with people moving to areas that are more inhabitable, resource shortages, and even wars associated with altered resource availability [88].

We clearly are altering the climate of the Earth in ways that will hurt much of humanity. Rising sea levels would threaten one tenth of the world's population, and we are already seeing the increased impacts of hurricanes on coastal communities already subject to sea level rise. Heat records continue to be shattered every year, and each heat record brings many related deaths. Over 1000 people die each year in the United States from heat-related causes, where there is much greater access to air conditioning than in many other countries. Climatologists linked the 2010 heat wave in Russia, in part, to climate change. It led to an estimated 15,000 deaths and cost their economy $15 billion. These are just small examples of the large-scale problems associated with increased temperatures. Crop yields could plummet, hazardous weather conditions become more and more common, superstorms become the norm, and a host of other changes occur. A combination of these effects could lead to crossing a threshold of a social tipping point. Global marine fisheries are already shifting because of warming, and their sustainability may be endangered, threatening food security for many [89]. The world will change in response to global climate change, and in many ways, we cannot yet anticipate.

Sunburn Revisited

Several billion years ago, cyanobacteria evolved a way to photosynthesize using water and giving off oxygen gas into the atmosphere. As this oxygen gas entered the atmosphere, it created an ozone layer in the stratosphere (6–30 miles up) that filtered out the harmful ultraviolet rays produced by the Sun. This allowed many organisms to evolve on land and in the sea, and the protective blanket of ozone remained in place for billions of years.

In the 1970s, production of refrigerants and propellants (chlorofluorocarbons) increased substantially. As these gasses enter the atmosphere, they spread to the stratosphere where they destroy the ozone. By the 1980s, there was a hole in the ozone opening over the South Pole. Scientists understood that the effects of increased UV would include greater rates of cancer, lower crop yields, and harm to most plants and animals on Earth. It was very clear humanity needed to do something and do it quickly to deal with the problem.

By the early 1990s, most countries agreed to control release of chlorofluorocarbons. This is a great example of global cooperation in the face of a potentially disastrous threat. Now, at least, the size of the hole that develops each year over Antarctica is not increasing and may even be recovering after 30 years [90]. Unfortunately, other processes still threaten stratospheric ozone.

Natural processes that act on nitrogen in the environment cause microorganisms to produce nitrous oxide from fertilizers that wash from cropland into the ocean. This nitrous oxide also destroys ozone. As we fertilize our world to grow ever more crops, the oceans receive increasing amounts of nitrogen and nitrous oxide production climbs. Nitrous oxide is also a greenhouse gas, and warming could increase the rates that oceans produce nitrous oxide. We are not out of the woods yet with regard to ozone destruction.

The links between global climate change, nitrogen fertilizers, and ozone destruction illustrate the complex and interactive relationship among global factors influencing human well-being and our collective actions. As we try to feed more people, we will need to fertilize more land. However, the greater use of fertilizer could lead to increased nitrous oxide, decreased upper-atmosphere ozone, increased UV, and ultimately decreased crop yields as the UV causes more crop damage. Controlling ozone depletion will clearly require global cooperation, more scientific information, and more clarity in global political goals.

Who Will Get Enough Water?

One out of eight people lack access to safe water. International disputes abound over water. Countries in dry areas are desperate for water to grow crops to feed their populations. These facts indicate how disproportionate the supply of clean water is around the world. Quality and quantity are both problems.

As I discussed previously, much water is too far away or comes in floods or quick snowmelt, and people simply cannot use it. Water is very heavy and costs a substantial amount of energy to move. Desalinating seawater is energy-intensive and a very expensive way to provide freshwater even when plenty of salt water is available. People already use half of the available water, and demand is increasing. The population is growing and using more water. Water pollution makes the water we have less useful and more expensive to use.

One of the large concerns is how changing climate will alter water availability. The warmer atmosphere will push more water into the atmosphere and lead to more precipitation worldwide. However, as heat increases, water evaporates more quickly as well, so it is not clear if plants (including crops) will benefit from the greater moisture in some areas.

Nearly 1/6 of the world's people rely on water supplied by snowmelt from the Tibetan Plateau and other snowy mountains. As the glaciers and snowfields melt, there is less certainly about where water will come from in the summer. A warmer Earth will lead to more snowmelt runoff in the winter and early spring, shifting it from the summer when demand is greatest [91].

All these facts lead experts to predict that water is one of the key resources for humanity and one that is most likely to run out. As we run out, the chance for societal conflict increases, although this conflict is difficult to predict and related to complex factors [92]. As we push the limits of our ecological footprint on the planet, water is one of the resources most likely to show exactly where those limits lie.

A World of Rats, Cockroaches, and Weeds

We are living in one of six great species extinction episodes in the 3.5 billion years that life has existed on our planet [93]. In the past, asteroids and perhaps volcanic activities have caused the extinctions. Now, we are responsible.

One of the main areas where people are causing extinction is in the tropical rain forests. Much of the diversity on land (plants and insects) occurs in these rain forests. However, people are cutting and burning forests very rapidly for lumber and agricultural production around the world. We cut about 39,000 square miles per year and degrade about the same amount. Roughly, half of all the tropical rain forests remain globally.

Another area where species are being lost is in the coral reefs. When carbon dioxide dissolves into seawater, it makes it more acidic. Acidic water makes it impossible for corals to deposit their projective shells, reproduce, and grow. As people increase the carbon dioxide in the atmosphere, the chemistry of the ocean makes it certain that this will lead to more acidic oceans. Given that corals have very narrow temperature limits for survival, global warming will have a further negative effect on these highly diverse ecosystems [94]. Marine habitats in addition to coral reefs are facing strong impacts on diversity due to habitat destruction and overfishing [95].

Larger animals are particularly vulnerable to overexploitation. Many of these large animals are predators. Removing the predators substantially changes the interactions of species they eat and all those on the food web below. Thus, we are changing food webs globally [96].

The trends all mean that over the next 50 years, half of all species on Earth will go extinct. While species extinctions are a normal and natural part of the long evolutionary history of our world, the current extinction rates are grossly elevated. Around 20% of birds and freshwater fishes are now extinct or endangered. Loss rates of freshwater animal diversity are among the greatest on Earth [97].

As we are causing extinctions, we are spreading pest species throughout the world. We move weeds, pest insects, diseases, and animals around the world at high rates. Some of these (e.g., trout) we move on purpose; others (e.g., rats) are stowaways. Our agricultural practices have allowed some species to become dominant over large areas on Earth (e.g., wheat, corn, rice, potatoes, and cassava). We encourage a few strains of these domestic plants to cover the planet while we allow extinction of related strains that could provide genetics to breed better pest resistance or yields.

The net result of extinctions and introductions is that the world is becoming more homogenous and less diverse. This combination of factors obviously interferes with the rights of species other than humans to continue to exist. The drastic alterations of species groups could also have other strong effects. Ecosystems are interconnected systems that upon occasion break down. We require the ecosystems of the Earth to regulate climate and provide the goods and services we need to survive. For example, pollinators are key to maintaining the plants in an area, including many of the food crops. If we cause extinction of pollinators, far more is lost than some insects and birds.

Conservation biologists have compared the loss of diversity to losing rivets on an airplane. If you are flying, and one of thousands of rivets that hold on the wing pops out, it may not matter much. Then if more go, it still might not matter. However, at some point that one last rivet that holds the plane wing on might pop off. We cannot predict exactly which rivet that will be, but we know that a breaking point might exist where the system stops behaving as it did and collapses [98].

One important set of "rivets" is the pollinators. Bees and other insects pollinate about a third of our fruit crops. Essentially all fruits and many nuts and vegetables require pollination. Surveys suggest that many species of bees no longer exist in areas where they previously did. Habitat destruction is partially to blame where conversion to cropland removes the forests where the bees nest. Another pressure is the widespread use of pesticides to control insect pests that also harms bees [99]. These chemicals can move into the pollen and nectar collected by bees. Global warming also threatens bee populations. Most species of pollinators depend on several species that flower at different times throughout the year. With global warming, the timing of flowering changes, disrupting the finely tuned evolutionary relationships that have developed over many thousands of years. This disruption is potentially leading to losses of species of both pollinators and flowering plants. Loss of these pollinators will lead to clear and quantifiable economic losses as well. Furthermore, wild pollinators provide about half the crop pollination. So using domestic bees to replace wild insects is not a viable strategy to replace lost pollinator diversity [100].

Protecting diversity will require conservation of large areas of land and ocean. These will need to be the high diversity areas. As the global climate warms, we may need to move many species to areas where they can survive or at least provide corridors for them to move naturally to those areas. We also need to control carbon dioxide emissions to halt acidification of the oceans and slow warming to save corals [101]. Controlling spread of invasive species will also help stem the loss of endangered species.

All of the steps to protect diversity will require local and global cooperation. The human race needs to decide if we collectively think other species have a right to continue to exist on the planet or if our needs trump those of all other species. We certainly need to preserve enough diversity to make sure the ecosystem that supports it does not collapse. If we do not, the world will be a poorer place for our descendants.

Up to Our Necks in Fertilizers

As we strive to grow more crops, the amount of fertilizer we add to those crops increases. We get nitrogen fertilizers by a chemical process that takes nitrogen gas from the atmosphere and converts it to forms that stimulate growth of plants (ammonia or nitrate). We mine and extract rocks with phosphate.

We have doubled the rates that nitrogen and phosphorus enter the natural environment. As the world population almost doubles, and expectations spread for higher-quality diets, we will need even more food. Growing this food will require expanding to lower-quality croplands and using ever more fertilizers. The rates of fertilizer use continue to grow, particularly in developing countries, and have increased to over 45% in the last 30 years [102].

As this fertilizer runs off, it causes many problems, several of which I mentioned already. The fertilizers cause algal blooms in lakes, leading to taste and odor problems and, in some cases, toxic blooms, costing society millions to billions of dollars each year [103]. Increases in nitrogen lead to greater rates of nitrous oxide entering the atmosphere, exacerbating both global warming and ozone depletion.

Nearshore areas of the oceans around the world commonly have dead zones. These zones form when fertilizer runs off the land and stimulates growth of algae in the ocean. When these floating algae die and sink, bacteria in the deeper water consume them and, in the process, consume oxygen dissolved in the water. The massive low-oxygen zones (the dead zone in the Gulf of Mexico off the Mississippi River can be greater than 5000 square miles) harm all marine life that cannot tolerate low oxygen and cannot get out of the way. The casualties include shrimp, crabs, shellfish, and fish. These massive fish die-offs harm economies and food webs in the oceans [104].

Dead zones occur all around the world including the Baltic Sea, the Black Sea, Chesapeake Bay, and coastal areas of Northern Europe, China, and Japan [105]. These dead zones will continue to form and become larger as more nutrient washes into the world's oceans. In addition to the dead zones in marine environments, the nutrients can stimulate algal blooms that are nuisance or harmful. In one internationally publicized incident, a large bloom off the coast of China impeded the Olympic sailing events, and the Chinese government spent a large amount of money and labor attempting to clean up the bloom. Nutrient inputs stimulate the toxic red tide that kills fish in oceans around the world.

As we attempt to feed a growing population, efforts to stem nutrient pollution will require increased technology. We will need to improve fertilization methods and ways we manage cropland to minimize nutrient runoff. Better sewage treatment and livestock management will also be necessary. Transfer of technology and information among countries will encourage improved fertilizer management.

Toxic Chemicals and Antibiotic Resistance

We are releasing chemicals into the environment globally without regard for the influences of many of these compounds. For example, agricultural pesticides are found in streams and lakes around the world at concentrations harmful to aquatic life [106] and potentially humans. Mercury contamination is in fish taken from distant areas of the oceans and all fish taken in areas where people live. Drugs, personal care products, and other compounds are put into domestic sewage, pass through treatment plants, and enter downstream waters. Many of these compounds are released until a problem is found with them (as opposed to human drugs where manufacturers must prove they don't cause a problem before they expose people to them).

Release of antibiotics into the environment is one example of global releases of chemicals that can have harmful effects on humanity. Heavy antibiotic use, including use to stimulate growth of otherwise healthy livestock, is leading to more and more antibiotic-resistant bacteria. Adding to the problem is the increased number of immune-compromised people that must continuously take antibiotics to control bacterial infection.

Over time, bacteria evolve resistance to antibiotics. They also have unique methods of sharing DNA, and the resistance can spread to bacteria that never cause disease as I discussed in Chap. 4. Worse, they can spread to other bacteria that do cause diseases, including tuberculosis. As such, tuberculosis cases are once again increasing in the developed world. Air samples taken in major cities around the world have antibiotic-resistant genes (DNA) in them, indicating the rampant movement of these genes is unavoidable if we continue to increase antibiotic use [107].

Currently, antibiotic-resistant bacteria are not the biggest problem in the world. Far more people die or suffer from common bacterial diseases that they could avoid if they only had access to clean drinking water or treatment with existing antibiotics. Still unrestricted and irresponsible use of antibiotics has led to more suffering in humanity than there should be given that controlling use of antibiotics is easy.

Paul Ehrlich once mentioned to me his worry about indiscriminate release of chemicals into the environment as discussed in Chap. 3. He remarked that we have no magic tweezers to get a chemical back out of the environment once it is released. His point is a good one. The bar for testing drugs is relatively high, but humanity releases hundreds of new chemicals into the environment for other uses every year.

Part of me is not worried about this, as people have had exposure to novel compounds synthesized by other organisms throughout our entire evolutionary history. No organism today can survive that is very sensitive to these unique compounds. On the other hand, people are synthesizing compounds that are even more unusual every day, the likes of which no organism has ever seen over the entire 3.5 billion years that life has existed and evolved on this planet.

For example, we now make many types of nanomaterials. These very small molecules or clusters of molecules have very unusual chemical properties. One of these materials is quantum dots. These are very small semiconductors with unusual electric capacities, such that they can store much more energy than any other molecules. We have synthesized very strange carbon compounds (tubes and spheres) that have properties that no other known molecules exhibit. We have embedded microscopic silver into many materials to inhibit bacterial growth.

The potential problem with these materials is that independent toxicologists have not rigorously tested them for their health and environmental effects. We have had issues like this in the past. In the 1950s in Minamata, Japan, there was a severe case of mercury poisoning with at least 1700 victims, probably more. The use of "agent orange" as a defoliant in the Vietnam War led to lasting health problems for soldiers and Vietnamese civilians that continue up to this day. Humanity has experienced accidental release of radioactive compounds that have caused problems for decades; the nuclear accident at Chernobyl has caused thousands of cases of cancer. Still, we release chemicals into the environment with minimal testing and/or standards in many parts of the world.

The fact that the amount of death and suffering related to such chemical releases is modest, at least so far, relative to some other causes does not mean that we should ignore the potential for problems. International regulation and cooperation in testing standards would be necessary to lower this potential risk from unforeseen effects of novel chemicals in the environment.

What Are the Causes and Solutions of the Largest Environmental Threats?

I list quite a few problems and their causes above. This chapter suggests problems mostly with respect to the viewpoint of humans. However, they also impinge on the rights of other species. Given all the causes, which are the overarching ones that are the most pressing? In 2019, the Intergovernmental Science-Policy Platform on Biodiversity and Ecosystems Services (IPBES) released their preliminary findings on the environment and changes over the last 50 years. This is an intergovernmental organization with members from more than 130 UN governments. Their initial preview of the global assessment found that the five biggest ultimate drivers of environmental damage are change in land and marine use, direct use of organisms, climate change, pollution, and invasive species. They considered how nature and ecosystem services have both declined.

This list is a reasonable start and clearly will take international cooperation to accomplish, though many of the specific causes have local origins. For example, controlling direct use of organisms requires local regulations and enforcement. However, as the efforts to control ivory trade illustrate, international pressure is an important ingredient for success in many cases. In some cases, international cooperation is essential to control damage. For example, land use in Antarctica and many marine areas is only subject to international law or agreements. We have a long way to go in controlling these problems. Control of human, corporate, and governmental behaviors is necessary to accomplish protection of the global environment. I will discuss these in more detail in Chaps. 11 and 12.

Chapter 8
By the Numbers: Ranking the Problem

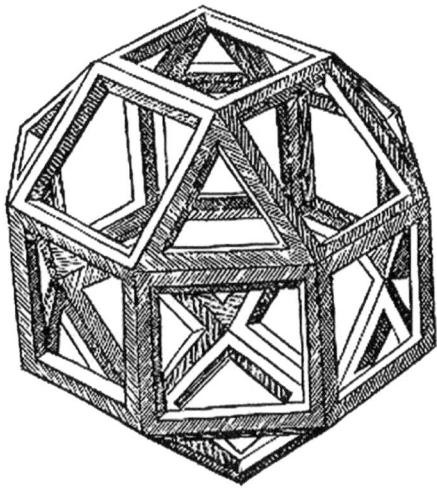

The first ever-printed version of the icosidodecahedron, by Leonardo da Vinci as it appeared in the Divina Proportione by Luca Pacioli 1509, Venise

In this chapter, I propose a way to rank the various problems on a relative scale and go through the estimates of death and mortality rates. Some of the estimates vary because of lack of data, and others are difficult, as they require predicting into the future. In these cases, I provide a range of estimates as a measure of uncertainty represented by high and low estimates of the index. This allows me to explore the sensitivity of the system to the assumptions made to create the ranking. I present both a "conservative" or optimistic estimate and a pessimistic estimate, the latter consisting of the direst predictions.

The original version of this chapter was revised. The correction to this chapter is available at
https://doi.org/10.1007/978-3-030-30410-2_14

While I have discussed many statistics so far, an overarching index to rank the relative threats or severity of the problems allows comparison of very different potential problems. I take a relatively simple mathematical approach. Jump to the rankings a few paragraphs down if this is too much for you; most people use indices without fully understanding exactly how they are calculated. Examples include Richter scale for earthquakes, F ratings for tornados, Saffir-Simpson Hurricane Wind Scale, stock market indices (e.g., the Dow, Nikkei), color-coded terrorist threat bulletins, and many others. Calculation details are given in Appendix I.

Fundamental Assumptions to Create an Index of World Problem Severity

I discussed the moral assumptions behind the book as a whole in the first chapter. These include that all people have equal value and the natural world has intrinsic rights. I do not know how to put the rights of the natural world into an index, so I will ignore that moral assumption in this chapter. However, many of the problems that directly harm humans also damage the planet. I needed to make several additional assumptions to create an index of world problem severity from estimates of current death rate, current suffering rate, future death rate, and future suffering rate.

Creating an overall index including both suffering and death is difficult. It might be comparing apples and oranges to equate current death rate and future suffering, but I have decided to do it based on several assumptions. Many people would (and do) choose to suffer for a long period rather than die. Does that mean a death should always count for more than suffering? Here I assume some long period of suffering is equivalent to one death, but I will explore how different weightings of suffering to death alter the rankings, to illustrate the sensitivity of the calculations.

Another problem is how to rank deaths and suffering. Is the absolute number of people suffering or the proportion of people suffering most important? For example, the percentage of undernourished people in the world dropped from about 19% to 11% between 1991 and 2017, and the absolute number went from 1.0 to 0.83 billion people. That is a 20% drop in numbers, but the proportion of people fell by 42%. Here I will use the proportion. The logic behind this is that the total annihilation of the human race does not seem worse if there are 1 billion or 7.5 billion people, and using the proportion of death and suffering signifies the general condition of humanity independent of global population at the time.

Using proportions also helps scale future suffering under a variable human population. In this book, I will use the number 7.5 billion people for current death and suffering calculations (roughly the global population in 2016). I will use the number of 11 billion people as global population for future death and suffering as this is the median estimate for global human population in the year 2100 [108].

While I give the specifics of calculating the index in Appendix I, I put it into words here. In this index, the potential death equivalents each year proportional to total population are the sum of proportional global deaths per year, potential deaths per year, suffering death equivalents, and potential suffering death equivalents.

Table 8.1 The index problem severity and the proportion of people dying in a year assuming a global population of 7.5 billion people and one of 11 billion

Value of PROBSEV	Proportion of global population dying	Number dying of 7.5 billion	Number dying of 11 billion
10	1	7,500,000,000	11,000,000,000
9	0.1	750,000,000	1,100,000,000
8	0.01	75,000,000	110,000,000
7	0.001	7,500,000	11,000,000
6	0.0001	750,000	1,100,000
5	0.00001	75,000	110,000
4	0.000001	7,500	11,000
3	0.0000001	750	1,100
2	0.00000001	75	110

Then the index transforms to a logarithmic scale to allow comparison of widely varied events.

I show how the index relates to the number and proportion of people dying in any 1 year in Table 8.1 to allow the reader to visualize the logarithmic nature of the index. Note that at a value of the problem severity index of 10, the entire human population is gone. At a value of 2, 75 people have died if the population is 7.5 billion, a number below the error of most global mortality estimates.

How to Compare Suffering and Death?

This question could depend on who is suffering and who is dying. The relative weighting of suffering and dying in practice varies widely depending on the situations; many people in the world are willing to allow the death of many other people so that they can be a little bit more comfortable. This is why world hunger and starvation remains a problem even though we have enough food to feed everyone. In contrast, the will to live and the optimism that things might be better in the future lead people to endure personal suffering for many years rather than dying. Yet, if there is a chance to end suffering (e.g., to feed your family for a year), many people are willing to risk death. This is how evolution has kept our species (and others) going. It is better to suffer and reproduce than not reproduce at all, but sometimes it is worth a chance of death to minimize suffering. Suffering is also relative. Some people would consider being without a cell phone suffering. I will consider the rough equivalent of chronic disease or hunger with the threat of malnutrition as suffering.

The variation in how people weight death and suffering leads me to create a range of values. On the low level, I assume 20 years of suffering equals a death. On the high end, I assume roughly two lifetimes of suffering (150) years is equivalent to one death.

World Hunger

I could not find solid estimates of the number of deaths per year due to starvation in the world. Several different organizations estimate the number of deaths of children under 5 years old caused by starvation. An interagency report [109] estimates that undernutrition was an underlying cause in 1/3 of the global annual childhood mortality which was 7.6 million in 2010. This translates to about 2.5 million deaths. The group Stop the Hunger [110] estimates about 8.3 million people each year die from hunger as of November 2018 or about 4 times more than the number of childhood deaths from the prior estimate. Thus, a range of 3–8 million people dying each year from undernutrition probably contains the true value and gives the range used for the calculation.

There are 925,000,000 people (slightly less than 1/6 of the world's human population) currently undernourished (according to the World Health Organization in 2010 [111]). There is some solidity in this number, as countries report the number of people in each country, their income ranges, and the total amount of food imported or produced (but not exactly how equally that food is distributed). The World Health Organization estimates the number of people who do not spend enough to insure adequate nutrition given data on how much each person spends on food each year. The Food and Agriculture Organization estimated that 821 million people suffered from chronic hunger in 2017 [112] and I count these people as suffering. If 1 in 100 of these people dies directly or indirectly from undernourishment, then the global estimate of mortality of 8 million deaths per year is reasonable. Part of the problem in determining numbers is that the reporting system for countries that are the most affected is the least reliable (because of infrastructure or unwillingness to admit the magnitude of the problem). The accounting for childhood deaths and overall malnutrition seems reasonably accurate however, and I do not explore a range of number of people suffering from hunger.

Global Nuclear Destruction

Global nuclear war has never occurred in the roughly 60years since the capacity has existed. The Cuban missile crisis in 1962 is, as far as we in the public know, the closest to nuclear war the world has been. The US Secretary of Defense at the time, Robert McNamara, stated that we were within a hairbreadth of complete nuclear war and humanity is lucky to have survived [113]. The United States and the USSR negotiated this international game of chicken safely, given the already clear potential for massive destruction. Somehow, humanity dodged that bullet.

The second potential trigger for all-out nuclear war is a world war involving the major nuclear nations. The last world war was over 60 years ago (perhaps because of the threat of total nuclear destruction). Between outright aggression and the potential for a mistake that leads to war, we have escaped this fate so far. A frightening number of technological false alarms have put us close to global nuclear war [114].

I assume there is between a one in a hundred and a one in a thousand chance that countries will detonate enough nuclear weapons to cause extinction of the human race because nuclear war has not happened since the USSR successfully tested its first nuclear weapon in 1949 (if we made it 70 years, we could make it 100). With current stockpiles, if Russia and the United States released even 1/10 of their nuclear arsenals, this would be enough to wipe us all out. Such an incident could be precipitated by errors (inadvertent release), sabotage (in particular computer hacking to cause perceived threats or really release weapons), a smaller-scale terrorist nuclear attack perceived as a full-out attack, or any of a number of other scenarios.

There is a greater chance of a limited nuclear war that will disrupt agriculture and cause death of 1/20 to 1/100 of the Earth's population in the first year and suffering for half the people on earth for 20 years. I will assume the probability of such an event is between 1 in 50 and 1 in 500. As more nations acquire nuclear weapons, this probability increases.

Finally, scientists have calculated the probability that high-energy physics experiments will cause global destruction [113]. Both the US Brookhaven National Laboratory and the CERN laboratory in Geneva calculated the probability that particle collision experiments would cause formation of strangelet particles (a hypothetical type of sub-atomic particle) and contagion of those particles. They estimated that if the experiment ran for 10 years, the chance of complete disaster for Earth was 1 in 50 million. As this is far less than other nuclear destruction probability scenarios, we will not include it in the potential for nuclear destruction. However, it does seem about as likely as destruction from large asteroids, megavolcanoes, and high-energy astronomic events.

Environmental Meltdown

It is extremely difficult for anybody to predict to what degree the global environment will fail given the current course humanity is on. The main problems lie in the ability to predict runaway greenhouse effects. Another is in how confident we can be that people will reverse some of most severe problems. Global tipping points, situations where we push the environment past a point to where it changes rapidly to an irreversible and undesirable state, are possible responses to human activities [115]. I will discuss such tipping points in more detail in the last chapter. On one hand, many countries are making no substantial effort to curb global warming; on the other hand, 196 countries signed the first globally ratified United Nations Treaty to control chlorofluorocarbon destruction of ozone. Several of the other potential problems (global nuclear war, collision with a meteor) would result in massive environmental destruction, but I do not consider them in this section as the death and destruction from environmental damage are accounted for separately.

I base my estimates on potential for death and suffering related to the global greenhouse effect and the potential for uncontrolled feedbacks that lead to massive warming. However, there are other potential negative scenarios such as running out of water in large portions of the world, running out of phosphorus fertilizers, and destruction of the upper atmospheric ozone related to excessive nitrogen fertilization (nitrous oxide releases).

In the conservative case, I will assume a 1/100 chance in any year that the environment fails to the point where half the people on earth suffer for 50 years. The optimistic case will assume a 1/1000 chance of such an event occurring. Overall, this is a relatively conservative calculation because most global environmental problems would take many more than 50 years to resolve, and it completely ignores the increased death toll that continues to grow for some environmental issues (e.g., with global warming far more heat-related deaths are expected [116], and hazardous weather is expected to increase). It also ignores some difficult to predict potential catastrophes such as release of a chemical that inadvertently causes an extreme drop in human fertility.

Meteor Impact

Estimates of the probability of meteor impacts are relatively consistent in the literature even though human history has not recorded world-damaging impacts. Estimates in the 1990s suggested a 1 in 10,000 chance that a 2 km wide asteroid or comet would collide with Earth killing a quarter of the humans through serious disruption of the ecosphere. The probability of a meteor that would kill most people on Earth similar to or larger than the one that probably caused the extinction of the dinosaurs is 1 in 100,000,000, and a collision that would result in the deaths of roughly a third of the human population would occur once every 500,000 years [117]. A more recent analysis suggests a continental scale impact would happen once every 100,000 years, a possible global catastrophe once every 700,000 years, and an object large enough to be certain to cause global catastrophe every 30,000,000 years [118].

Larger objects are easier to detect and, presumably as technology improves, become even easier to detect and would be easier to deflect over time. Here I assume an object with a 500,000-year return time would kill 1/3 of population and cause the remaining 2/3 of the population to suffer for 20 years before the infrastructure for food production and medical care is restored as the optimistic scenario. The pessimistic scenario is a "dinosaur killer" asteroid at 1 in 100 million that kills 95% of the people and the remaining 5% suffer for 20 years.

Magnetic Reversal

I assume no deaths would occur with magnetic reversals, but the disruption of electronics would occur. This approach probably misses some deaths that actually would occur; severe electronic disruption could cause airplane crashes, failure of medical equipment, and many other deaths, but these possibilities will be ignored here. The last short-lived reversal happened about 35,000 years ago [119]. I assume 1/100–1/1000 of the people on Earth would suffer for 1 year with the global disruption of commerce that could result from disruption of electronics as the pessimistic and optimistic scenarios.

Super Volcano

I assume that a volcano large enough to disrupt agriculture seriously will do so for 20 years and that half the people on Earth will suffer for this length of time. The probability of this large volcanic eruption is considered to be once every 0.1 million years or more [120]; for our purposes we will use this as a minimum and estimate chances at 1 in 0.1 million to 1 in 10 million. The worst-case scenario is an eruption or series of eruptions—the size thought to have caused the Permian-Triassic extinctions. As this occurred 250 million years ago, 1 in 250 million is the chance for the most catastrophic megavolcano. I assume that a large volcano will cause suffering for 10% of the people on Earth for 10 years and that the megavolcano will cause suffering of half the people for 20 years. I assume that a negligible proportion of the global population will die directly under either scenario.

Gamma Ray Burst

There is a chance a long gamma ray burst with a dose of 100 kJ/m^2 has hit the Earth in its history and led to a global extinction event. The probabilities are 90% over 5 billion years, 60% over 1 billion years, and 50% over a half billion years [121]. I assume an "extinction event" will cause death of half of the people and lead to suffering of the other half for 20 years, with an optimistic chance of 1 in 5 billion and a pessimistic chance of 1 in a billion.

Current Infectious Disease

According to the World Health Organization in 2015–2017 from the Global Health Observatory data repository [122], of the top causes of communicable diseases, in millions, about 2.9 (million) people die each year from respiratory infections, 1.4 from diarrheal disease, 1.02 from HIV/AIDS, 1.3 from tuberculosis, 0.4 from malaria, and 0.2 from hepatitis. Influenza alone kills between 291,000 and 646,000 people each year [123], and many more suffer from this mostly preventable disease. All causes sum to 8.6 million people.

   The number of people infected by the major diseases is at least 2 billion. This includes, in millions, 10.4 by tuberculosis, 36.9 by HIV, 212 by malaria, and 257 by hepatitis. Additionally, 1.6 billion people in 2015 required treatment for neglected tropical diseases not on this list, many of which are preventable [122]. New infections each year also include diarrhea (66 million), upper respiratory infections (235 million), meningitis (14 million), tetanus (0.16 million), measles (0.245 million), and whooping cough (1.5 million) [124].

   All the numbers could be low as many countries either do not report data or are not able to ascertain true rates in rural areas. In addition, some deaths may have more than one cause (e.g., malnutrition and multiple diseases could lead to mortalities). Given that the uncertainties associated with influenza alone range about two-fold, I assume annual mortality ranges from 5 to 11 million per year. The confidence intervals reported for the communicable diseases by the Institute for Health Metrics and Evaluation are generally tighter than 2-fold and run from 1.5-fold to 50%. I will assume the higher of the numbers and an optimistic number of 1.5 billion and a pessimistic number of 2.5 billion people per year suffering from preventable communicable diseases.

Pandemic

A pandemic that kills a significant portion of the Earth's population occurs every few hundred years. The fact that we can treat diseases offsets the increased probability of new diseases with greater population density, denser livestock production, and more contact with wildlife. The "Spanish" influenza killed about 3% of the world population in 1918–1920 [125]. The Black Death Plague in the 1300s killed around 50–100 million people [126] and reduced global population from 14% to 28%. Thus, I assume there is between a 1 in 100 and 1 in 500 chance that a disease will cause the death of one fifth of the population on Earth and lead to suffering of another fifth of the population for a year.

Caveats

A first criticism could be of my estimates of future probabilities of future death and suffering. Some of my guesses are admittedly just that. I try to base them on as much information as possible, but some of them, I simply needed to guess based on past experience and provided broad ranges of potential harm.

   A second criticism of my approach to calculation of a problem index could be that I do not weight the rankings by age. There are good reasons to consider age in the calculations. Older people have less life left, less time to suffer, and less to receive from and provide back to society. For simplicity's sake, I did not age-adjust my estimates. The issues that disproportionately influence children (hunger and disease) occur in developing countries with high populations and an age structure with relatively more children. Hunger and disease estimates would increase if I weighted death and suffering of younger people more heavily with an age-adjusted approach. As these problems are already among the worst with the current ranking system, it would not change the relative results much anyway.

A third major problem with my ranking approach is that it completely avoids the problem of damage to species other than humans. As we cause the complete extinction of other species on a regular basis, humans are the worst problem on Earth for most species of animals and plants on the planet. However, it is difficult enough to rank suffering against human death. How to rank human problems against existence of other species is even more difficult, and I will not deal with that problem here.

A fourth major problem is the juxtaposition of problems that are happening now against risk analysis for future events. We are certain people are starving and dying of disease today. There is far less certainty on how human behavior may or may not lead to nuclear war or if a large comet that has not been charted will slam into Earth. It is still of interest, to me at least, to contrast these problems using the common currency of potential and actual human death and suffering.

A fifth problem is trying to separate problems. For example, current malnutrition is a large contributor to deaths from diseases. Increased pressure to feed more people will lead to more production of cropland causing greater environmental damage. All of these issues will increase the probability of societal breakdown.

My best hope for the general approach here is not that my particular ideas will get traction; rather my hope is that the general approach will generate thought and discussion about problems at a global scale and their solutions. So in spite of the caveats, I will be successful if this causes you to think more deeply about our shared global problems.

The Estimates and Rankings

Finally, to the heart of the matter. While the graphs tell the story, I also realize that some people do not do well with graphs and prefer verbal information. For those looking at the graphs, keep in mind that I plotted them on log scales, so a bar that is twice as high visually is actually many times higher. Thus, I will discuss the major trends in the graphs of the results. In general, the nonhuman-caused problems (gamma ray burst, magnetic reversal, super volcano, and meteor impact) are less severe than the human-caused problems.

World hunger and current disease are the top worries with respect to deaths now and suffering now (Fig. 8.1). The variation in the deaths now occurs because my optimistic and pessimistic scenarios vary. The variation in suffering arises because I explore the two assumptions that one death is equivalent of 20 or 150 years of suffering. With respect to potential deaths, the top problems are global nuclear war, pandemics, and limited nuclear war. Potential suffering is greatest in the categories of environmental meltdown, limited nuclear war, and pandemics (Fig. 8.2). There is

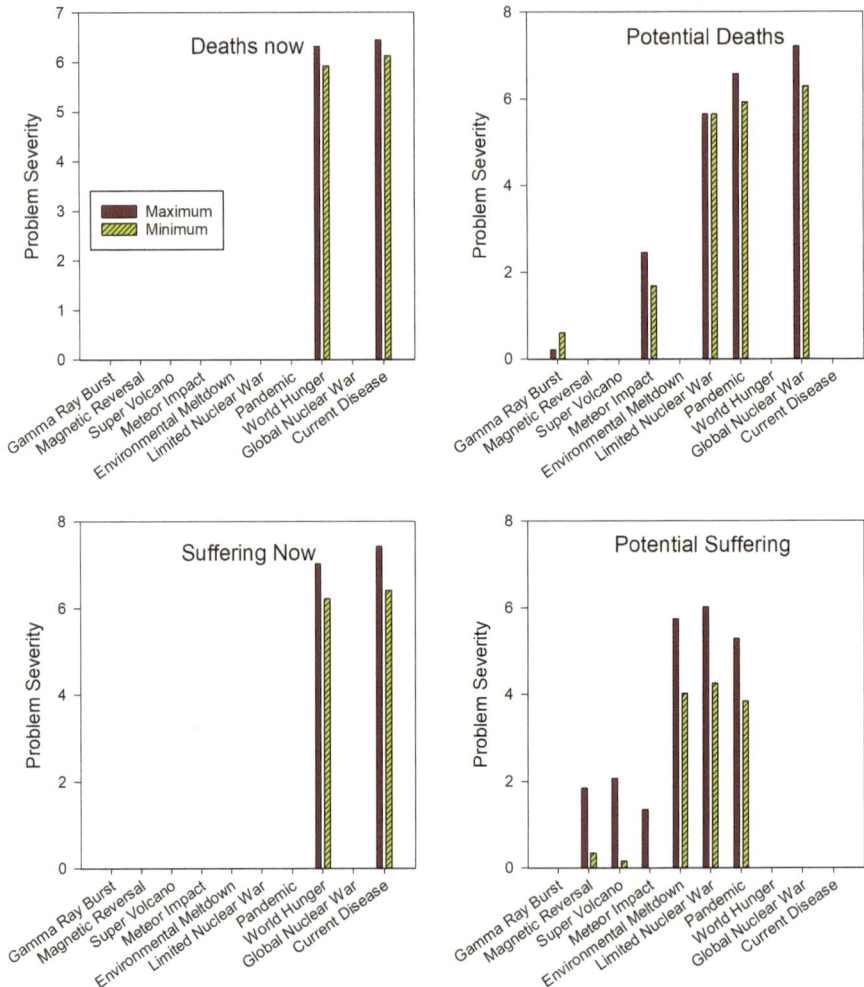

Fig. 8.1 Severity of problems broken into current and potential future death and suffering. The ranges in estimates represented by the paired bars are generated by optimistic (minimum, best-case scenario and one death per year = 150 years of suffering) and pessimistic (maximum, worst-case scenario and one death per year = 20 years of suffering) views. Please note the Problem Severity scale is logarithmic so bars do not sum

no potential for future suffering in global nuclear war because I assume humans will mostly be extinct within a year in this scenario. Variability in these estimates is driven in the same way as it is for deaths and suffering now.

The nonhuman-caused problems have modest impacts on future death and suffering, not so low that they should not be worried about at all, but not nearly as controllable or likely to be as devastating as the human-caused problems.

If we look at the overall summation of the problems (Fig. 8.3), nuclear war in general (and particularly if we consider total global war and limited nuclear war together) is an extremely worrisome problem. Note that two problems that are

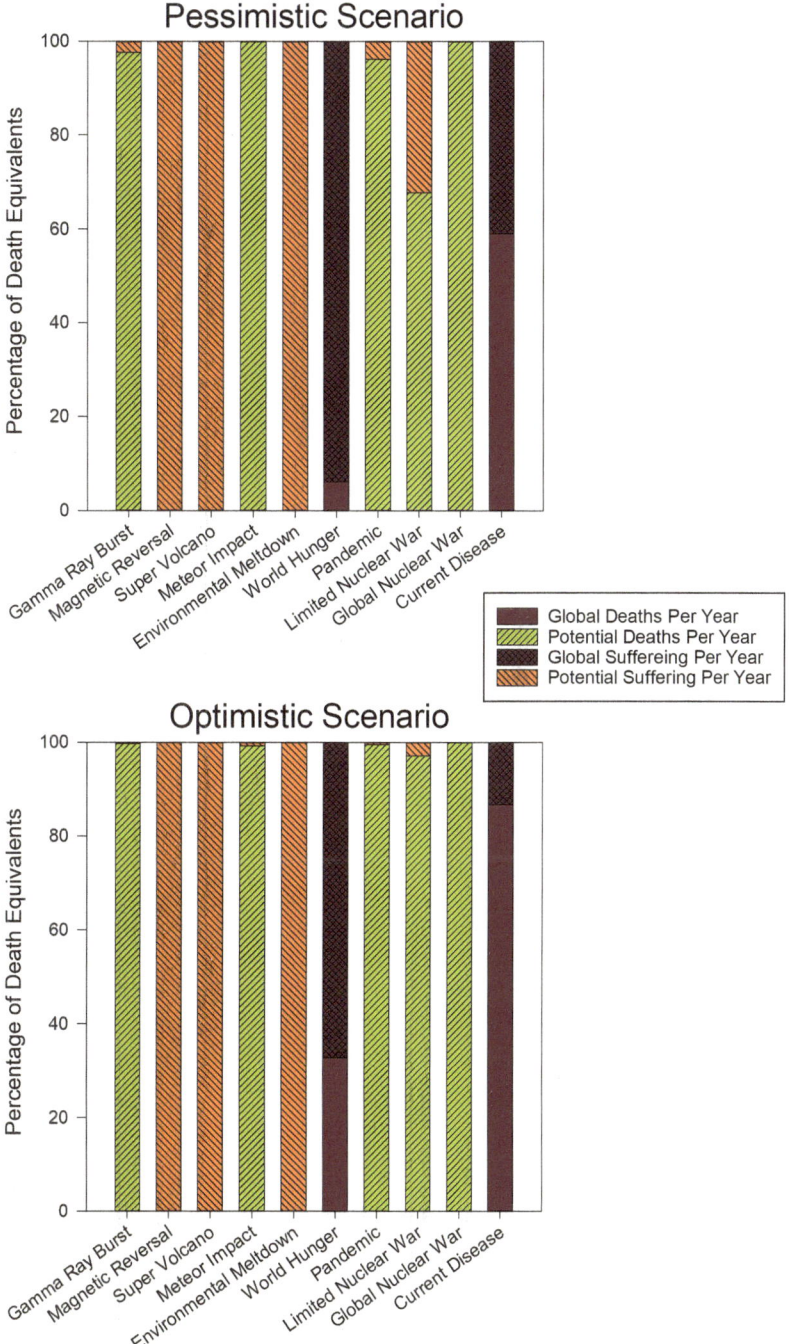

Fig. 8.2 Percentage contributions of deaths now, suffering now, future deaths, and future suffering to pessimistic (worst case and 20 years suffering = one death) and optimistic (best case and 150 years suffering = one death) scenario to the overall index

Fig. 8.3 Overall problem severity as calculated by summing current death and suffering from each cause and potential death and suffering. The range between maximum and minimum estimates is generated by optimistic and pessimistic estimations of threats as well as how death is equated to suffering as explained in the text

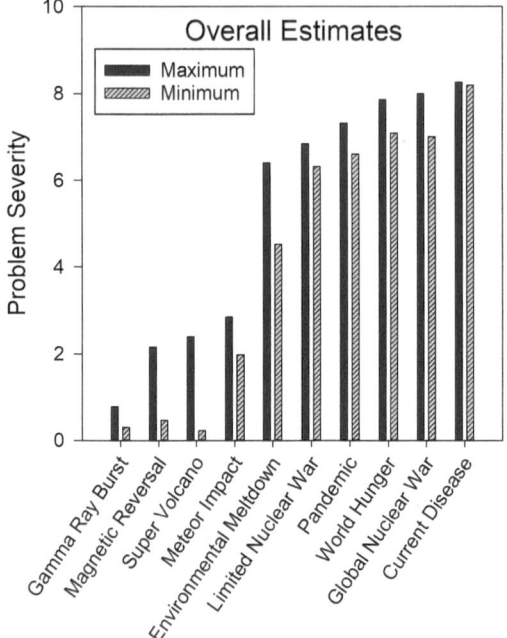

roughly the same magnitude added together double the estimate of the problem, but each unit on the scale is a tenfold increase. It makes sense to consider these two problems together, because the chief cause, nuclear weapons, leads to both issues.

Disease is also an important global problem; I consider future pandemics separately from current disease. This is because much current death by disease is preventable and predictable. Providing sanitary food and water, vaccinations, and general access to medical care could drastically lower these rates. In contrast, preventing pandemics requires increasing capacity for rapid response to novel disease agents, plans to limit spread of novel emerging diseases, and measures to minimize moving of diseases from nonhuman animal populations into human populations.

Global environmental meltdown has the potential to cause considerable suffering in the future, but the uncertainty is greater over this than some of the other human-caused problems, so the variation is wider in the estimate.

Chapter 9
What Do Other People Think the Worst Problems in the World Are?

DISCUSSING THE WAR IN A PARIS CAFE.
SEE PAGE 304.

Discussing the War in a Paris Café—a scene from the brief interim between the Battle of Sedan and Siege of Paris during the Franco-Prussian War Illustrated London News 17 September 1870

© Springer Nature Switzerland AG 2019
W. Dodds, *The World's Worst Problems*,
https://doi.org/10.1007/978-3-030-30410-2_9

Informal Surveys

I have asked numerous people the question, "What do you think the worst problems on Earth are?". Some people answer with causes of problems and other with effects. This is a very interesting exercise because it reveals the first answers that people have and has given me insight into some problems I never would have considered. I stress that I am shortening and paraphrasing the answers here; these are not direct quotes. This is just a selection of the answers to give an idea of how differently people approach the same question (Table 9.1).

During fall semester 2011, I asked the 80 students in an Introductory Biology class to answer the question "What are the worst problems in the world". This was an extra credit question on a quiz. I summarize the results on Table 9.2. Some students focused on causes of problems (more ultimate problems) and others on effects (proximate results). Students could give more than one answer, and some put both causes and effects into their answers. Hunger, war, poverty, and environmental issues ranked high. Interestingly, this survey occurred during a time when the economy in the United States was a central political issue. Accordingly, economy was at the top of the causes list. An approximately equal number assigned religion and lack of ethics/morality as a cause. Some problems I consider in this book did not even make it onto the list.

In 2014, I visited the University of São Paulo and polled students on what they thought the world's worst problems were; the responses were somewhat similar (Table 9.3) to those of the American students in 2011. In 2016, I asked the same question of an audience at the Federal University of Bahia, Brazil. The responses remained similar (one person responded that the worst problem is bad questions, which was a pretty funny answer). The responses make it clear that there are many viewpoints on what are the worst problems.

Jared Diamond has spent much time thinking about the problems that face humanity; in his book *Upheaval: Turning Points for Nations in Crisis* [127], he lists the worst problems: nuclear weapons, global climate change, global resource depletion, and global inequalities. He does not give rationale for these choices. Other than nuclear weapons and climate change, these are difficult threats to classify under my methodology as two hinge on social conditions ultimately leading to negative consequences for people (death and suffering), but I do not know how to relate these consequences in probabilistic terms as done with my comparative index.

Professional Surveys

More comprehensive scientific polling procedures give rather similar results. The European Union polls all their member countries on a variety of issues. In one survey, they asked "Which of the following do you consider to be the single most serious problem facing the world as a whole?" and gave a list of choices. They

Table 9.1 Informal responses to the question "What do you think are the worst problems on Earth?"

Respondent	Response
Adam Gussow	English professor and harmonica player: there is an imbalance between techne´ and Gaia: we're tearing up the natural world in what will surely, in the long run, prove to be an unsustainable way
Bart Grudzinski	Graduate student: people do not understand the consequences of their actions
Chuck Rice	Soil scientist: poverty
Claire Ruffing	Graduate student: access to freshwater
Dustin Shaw	College student: the rate humans use natural resources
Heather Brink	College student: energy conservation
James Tiedje	Microbiologist: global change, population
John Oehlert	College student: lack of understanding that we all share one planet
Katie Costigan	Graduate student: lack of education, particularly for women
Kendra McLauchlan	Geographer: poverty
Paul Ehrlich	Evolutionary ecologist: global climate change, pandemics, and nuclear war. Also, release of untested chemicals in the environment.
Rita Colwell	Environmental microbiologist: lack of education for females in developing countries
Russel Wohler	College student: ignorance
Zachary Cordes	College student: overpopulation and overconsumption

surveyed 26,840 people in 2011 and found that these people thought poverty was the biggest problem (Fig. 9.1). The same survey in 2009 had somewhat different results, suggesting that public opinion on what are the worst problems in the world changes quickly.

Another survey of young people around the world contained interesting results. The Global Shapers Survey is a project of the World Economic Forum to establish the concerns of young people around the world for policy makers to consider (http://www.shaperssurvey2017.org/). The project considers the fact that young people have little input on policy and government actions. In 2017, they interviewed 31,495 people from 186 nations around the world. The survey was available online in 14 languages including all official languages of the United Nations. Questions included: "In your opinion, what are the most serious issues facing the world today?" Possible answers were (1) lack of education, (2) food and water security, (3) lack of healthcare services, (4) large-scale conflict/wars, (5) lack of economic opportunity and employment, (6) inequality (income, discrimination), (7) government accountability and transparency/corruption, (8) lack of political freedom/political instability, (9) climate change/destruction of nature, (10) loss of privacy/security due to technology (online privacy/cybercrime/social media trolling), (11)

Table 9.2 Synopsis of results to the question "What do you think are the worst problems on Earth?" as asked to 80 introductory college biology students

Effects	#	Causes	#
Hunger	16	Economy	6
War	11	Selfishness	5
Poverty	10	Resource use rates	5
Pollution	9	Greed	5
Biodiversity loss/deforestation	6	Overpopulation	5
Global warming	5	Loss of ethics/morality	4
Disease	5	Religion	3
Natural disasters	3	Hate	2
Lack of universal health care	2	Lack of education	2
Agriculture problems	1	Energy sustainability	2
Sex trafficking	1	Economic disparity	2
Fatherless homes	1	Prejudice	2
Loss of agricultural land	1	Human effect on environment	1
		Politicians not helping us	1
		Racism	1
		Violence	1
		Lack of technical solutions	1
		Waste/inefficiency	1

Table 9.3 Synopsis of results to the question "What do you think are the worst problems on Earth?" to a group of mostly college students at the University of São Paulo

Problem	#
Poverty/inequity	16
Mismatch between human logic and nature/politics/lack of love in human heart/not in harmony with nature/greed	14
Education	11
Greenhouse effect/non-sustainable actions/environment	9
Population growth/people	8
Crime/violence	6
Hunger	5
Disease/health	4
Disasters/people in wrong place/accidents	3
Wars	3
Political instability	2
Food production	2
Economic growth, resource exploitation/water limitation	2
Cheap goods	1

religious conflicts, (12) poverty, safety/security/well-being, (13) lack of infrastructure, (14) ageing population, and (15) others. The top answers are in Fig. 9.2. These answers roughly align with my nonscientific small surveys and the European survey.

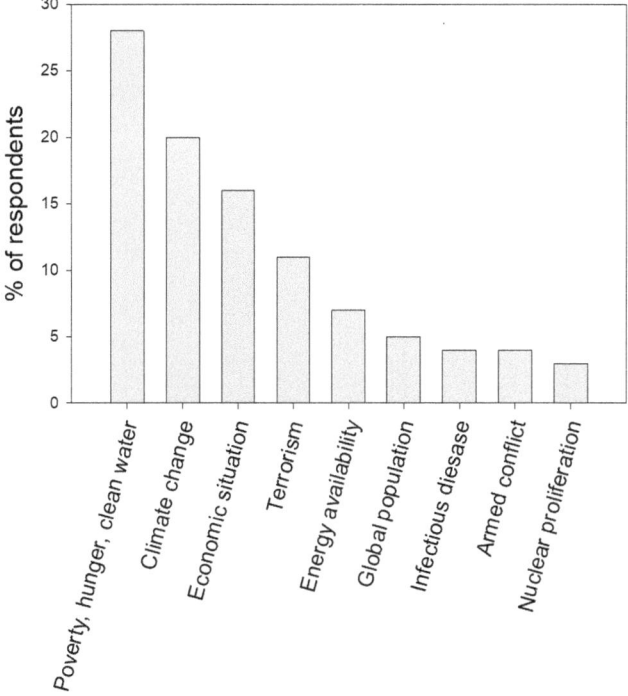

Fig. 9.1 Survey of Europeans asking "which of the following do you think are the worst problems?" (http://ec.europa.eu/commfrontoffice/publicopinion/archives/ebs/ebs_372_en.pdf)

Another initiative to determine the worst problems on Earth seeks to help young people choose careers in which they can make the biggest positive effect. This approach ranks the top ten problems differently, based on resources necessary to solve a problem, how crowded the field is (number of people working on the problem), and problem severity. The project is named "80,000 hours" (referring to the total numbers of hours that people work in a career) and combines research from the Open Philanthropy Project, the University of Oxford's Future of Humanity Institute, the Copenhagen Consensus Center, and that done by other groups.

This project lists the top ten problems as of 2017 in decreasing importance as (1) risks from artificial intelligence, (2) promoting effective altruism, (3) global priorities research, (4) factory farming, (5) biosecurity, (6) nuclear security, (7) developing world health, (8) climate change (extreme risks), (9) land use reform, and (10) smoking in the developing world. As you can see, some of these are substantially different from those already considered, but the goal of this list is different from some others. This project seeks to identify areas where young adults can enter a career that will be likely to help solve a large-scale pressing problem.

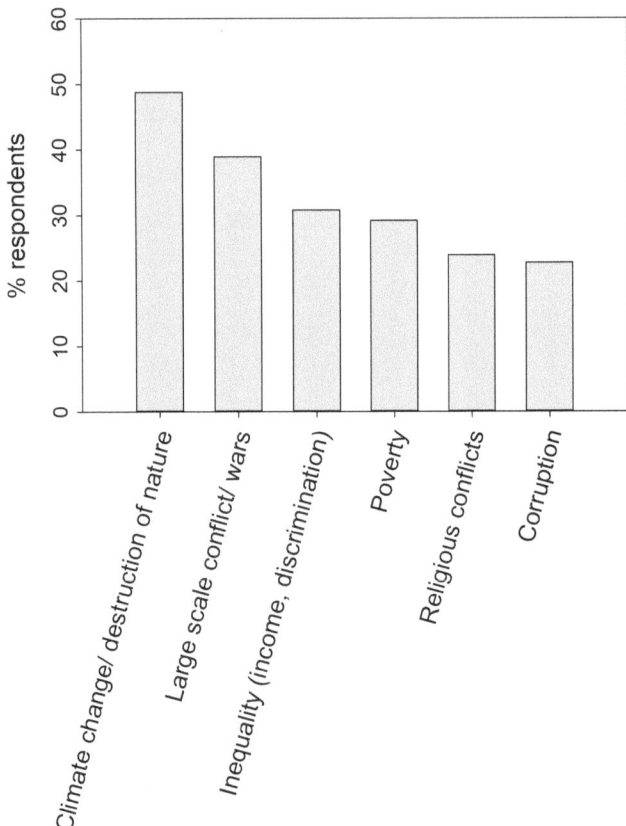

Fig. 9.2 Results of global survey of people between 18 and 35 years old in answer to the question "In your opinion, what are the most serious issues affecting the world today?" (http://www. shaperssurvey2017.org/static/data/WEF_GSC_Annual_Survey_2017.pdf). Respondents were asked to pick three responses so totals come to more than 100%

People's Thoughts on the Worst Problems Is a Function of Current Events and Local Conditions

A Pew center 2014 survey (Fig. 9.3) on what people thought were the worst threats in the world indicated substantial variation across regions in some issues, and others were relatively constant. For example, religious and ethnic hatred were seen as the worst problem by over 1/3 of the people surveyed from the Middle East (an area with substantial turmoil related to religious unrest). People in Europe and the United States saw inequality as the worst problem. Latin Americans viewed pollution and environmental destruction as worst, and Africans saw diseases as the worst problem. There was substantial variation regionally in these areas. In contrast, nuclear weapons were a severe problem for close to 25% of people regardless of region. More locally, 49% of Japanese saw spread of nuclear weapons as the worst problem. This makes sense, as this is the only country ever directly attacked by nuclear

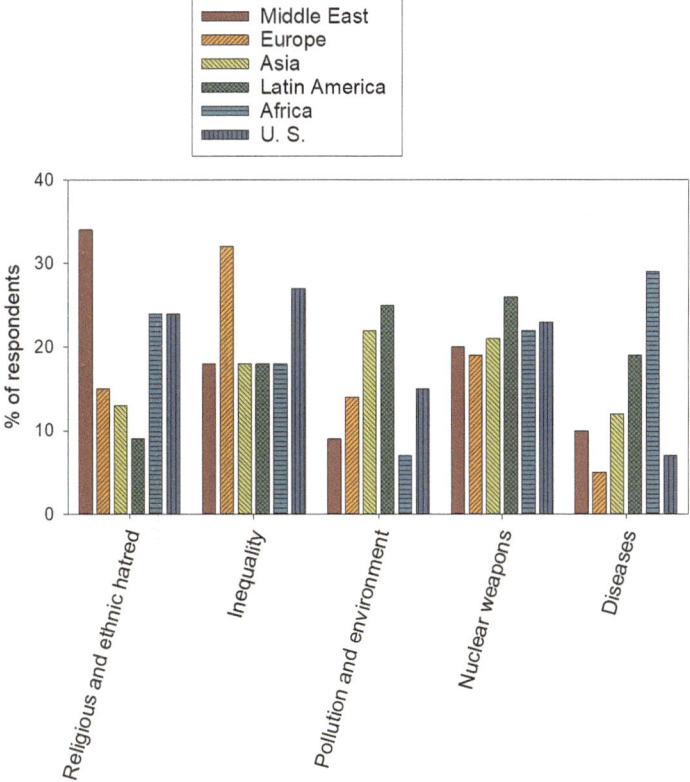

Fig. 9.3 Results of a Pew center 2014 survey on global opinions of worst problems broken up by regions of the world. http://assets.pewresearch.org/wp-content/uploads/sites/2/2014/10/Pew-Research-Center-Dangers-Report-FINAL-October-16-2014.pdf

weapons. At the time of the survey, not only was AIDS a widespread problem, but the Ebola epidemic was recent. In Nigeria, where Boko Haram terrorists were very active, 38% thought religious and ethnic hatred were the worst problems.

This survey was a repeat of a survey from 2007. The main shifts, globally, were an increase in the proportion of people who thought that religious and ethnic hatred was the worst problem and decreases in the proportion of people who thought diseases and environmental problems were the worst. These surveys suggest that current events locally shape concern about global problems.

Is Opinion more Important than Objective Ranking?

The worst problems are whatever the most people think they are. If more people in the world think a problem is the worst, then that is the worst problem. This has its attractions; solving a problem could be easier if most people agree the problem needs solving. It also has its pitfalls, because opinion and fact are not necessarily the same. One problem with the surveys I discussed in this chapter is that they are not open-ended. The people who wrote the surveys started with a list of what they thought the categories should be. I described some informal open-ended surveys and discussions in the first section of this chapter. The open-ended responses gave a much wider array of answers. In some ways this is more reflective of what people think are the worst problems. The issue with the survey approach and translating it to action is the answer might change from day to day, even for the same person.

I think that overall, these responses and surveys indicate that people do think globally (although there are local differences in what people think are the worst global problems) and that some of the issues they think are important line up well with the rankings that I present based on an objective index. However, many of the problems, particularly those that deal with social unrest, rely upon assessing the probability such unrest will happen in the future. Groups seem to do better than individuals at making predictions (according to social science). When asking what the worst problems are, many of them (nuclear war, terrorism, environmental damage) require prediction of the future severity of the problem. However, in all cases, accurate information improves prediction, and the process of creating the index I outline here provides information and allows separation of information from moral choices.

Chapter 10
Progress Toward Solving the Problems and Potential Costs of Solutions

Illustration to Archimedes "Give me but one firm spot on which to stand, and I will move the earth" (Oxford Dictionary of Quotations, 1953) Engraving from Mechanic's Magazine

Solving big problems requires political will and money. The less money required, the easier it is for political systems to work to solve these problems. Some problems (e.g., life-extinguishing gamma ray burst) have no solution and so cannot be solved by any amount of money. For other problems, the data are just not there to make cost estimates. However, here I try to make some estimates.

The generally agreed method to estimate global economic activity is the global gross domestic product. We can calculate gross domestic product in several ways. The simplest calculation is the total corporate and individual income earned. The World Bank estimated, in 2017, that the global gross domestic product was $81 trillion; the United States was responsible for $19.3 trillion of this. The global gross domestic product has grown substantially over the last 25 years. If the gross domestic product per capita is calculated, there is about 2% per year increase over the last

© Springer Nature Switzerland AG 2019
W. Dodds, *The World's Worst Problems*,
https://doi.org/10.1007/978-3-030-30410-2_10

100 years, with some bumps and dips. Given these numbers, $1–2 trillion per year (the average economic growth) to solve the world's biggest problems does not seem unreasonable given that the global gross domestic product is increasing by about that much each year. Solving the problems is insurance for the economic system as a whole, as many of the problems I have covered here have substantial economic ramifications. Some solutions, such as controlling disease and hunger, could increase economic overall productivity and growth.

Additional limits include the energy required to solve problems such as poverty and public health [128]. Countering these limits are vast increases in technology to solve problems. For example, remote sensed satellite data can be used to predict areas where poverty is most prevalent, and these data can be used to target efforts to counter the poverty in the areas where data are most difficult to come by [129]. One of the biggest technological advances that have improved the lives of people in developing countries is the spread of cell phones. My trips to Mongolia and Panama made me realize that many people in remote areas now have connection to the outside world because installing cell towers is so much less expensive than were telephone lines.

Costs to End World Hunger

The United Nations estimates it would take $30–50 billion per year to end world hunger. This is less than one tenth of 1% of the global gross domestic product and less than 1/3 of economic growth each year. People have made much progress in fighting hunger at the global level. According to the UN World Food Program, the absolute number of people is 842 million who do not have enough food to eat, a 17% drop since 1990. This represents a change from 19% of humans suffering from hunger in 1990 to about 12% in 2013. The World Food Program calculated in 2012 that $3.2 billion per year would be required to reach all 66 million hungry school-age children. Assuming that it costs twice as much to feed hungry adults, then about $40 billion per year could feed all the people. Remember, total global gross domestic product is about 2000 times more than $40 billion.

Deeper solutions to hunger and disease include education of all. While education of all people would cost substantially, the economic stimulus would far outweigh the costs. Most experts agree that education stimulates economies, and one report estimates that the global economy loses $1 trillion per year in lost income due to illiteracy [130]. Education, particularly of women, allows more people to enter the workforce, leads to smaller families, and increases income.

Costs to Protect the Environment

Protecting a substantial portion of Earth's species will cost $76 billion per year [131]. This is roughly one tenth of a percent of the global gross domestic product $(0.001 * GGDP)$. Thus, solving this problem should be well within the reach of humanity. However, there is little incentive to protect species. Over half the loss of biodiversity is due to heavy resource use in some countries causing biodiversity loss elsewhere [132]; thus there is little incentive within countries to preserve biodiversity.

Global warming caused by release of greenhouse gasses is a difficult problem to solve. Economic activity correlates directly to rates of fossil fuel consumption. If we discover low pollution solutions, such as fusion reaction (controlling miniature suns on Earth) or highly efficient solar collectors, then the cost will not be too great. Currently, William Nordhaus, a prominent economist from Yale University, member of the US National Academy of Sciences, and recent Nobel Prize winner, estimates that the world would need to spend 2–4% of income to control global warming. Much of this control will revolve around curbing resource use in affluent countries [133], which may not prove popular in these countries. While this may seem a substantial amount, it is well within the variation caused by the vagaries of economic fluctuation that occur anyway.

Part of evaluating the costs of protecting the environment against the benefits of protecting it may lie in properly accounting for the ecological services the environment provides. Just as people regularly check the gross domestic product or stock market indices, an accounting for the value of ecosystem goods and services and how that value changes over time could assist in managing environmental exploitation and protection. This could increase the benefits that the environment could provide for humanity [134].

Costs to Eradicate Disease

Eradicating polio globally could cost several billion dollars, yet the annual economic costs (not even considering the suffering) of the disease have exceeded that over decades [135]. The cost of globally eradicating any disease is a onetime, rather than a recurring, cost, and even local eradication efforts can stop a huge amount of suffering and costs associated with diseases.

Controlling (perhaps eliminating) malaria globally could cost $5 billion per year according to the Global Malaria Action Plan. According to the World Health Organization, current international spending, including domestic and international expenditures, is $2.5 billion per year. These costs would decline as the prevalence of malaria decreased. UNICEF estimates that malaria costs Africa alone $12 billion per year. The World Health Organization spends almost $2 billion each year on reducing all diseases, and this does not include financial input from individual countries. Eradication of all diseases that we have the technology to eliminate, and control of those that are chronic could well cost hundreds of billions of dollars each year, but the costs of diseases to society are considerably greater than that.

According to data from the World Health Organization, the global number of child (0–5) deaths fell by about 1/3 from 1990 to 2012, much of this related to decreases in disease. Over the last decade, there was a 30% decrease in pneumonia and malaria, a 50% decrease in diarrhea and AIDS, and an 80% decrease in measles [136]. Unfortunately, decreases in immunization rates in some developed countries are undoing some of this progress [137].

Some of the steps to decrease death and suffering from disease are simple. In 2000, the United Nations projected it would cost $10 billion to provide all people on Earth safe water and noted that the number is "one-tenth of what Europe spends on alcoholic drinks each year, about the same as Europe spends on ice cream and half of what the United States spends each year on pet food." So, while the number sounds large, relative to many global-level costs, it is not that great. The report also noted the political hurdles to solving the problems are probably as great as challenges of actually raising the money.

The UN reported in 2012 that "Between 1990 and 2010, over two billion people gained access to improved drinking water," meaning that the number of people without access to clean water was cut in half. We continue to make great strides in providing sanitary water, and this is a truly bright spot. Still many do not have access to clean water, and waterborne illnesses are one of the major causes of global death and suffering.

Costs to Protect Against Asteroids

Costs range from detection of the object (space observation) to plans to divert the object and civil disaster preparation [138]. The United States National Research Council estimates roughly $50 million per year could detect most potentially dangerous asteroids, and $250 million per year could support further research and a space mission for deflection capacities. A private foundation has the goal of helping find asteroids that could threaten Earth. They estimate that it will take $450 million to protect Earth. This cost is to put up an infrared (light in an invisible part of the spectrum that we sense as heat) telescope on a space station parked near Venus. This group was created because of concerns that NASA and other international governmental space agencies are not doing enough to protect humanity from collisions with objects from space. However, international cooperation is strong in the astronomic community, and multiple space agencies worldwide are developing complementary missions to deflect asteroids [139], indicating willingness to spend resources to resolve the problem.

Costs to Protect Against Global Disruptions to Food Supply

Storing food would be necessary to deal with the major disruptions to the world supply. These include nuclear winter, meteor impacts, partial gamma ray exposure, societal collapse, and super volcanoes. Estimates are that there are 4–7 months of food stockpiled globally [140]. However, years' worth of food would need to be stored for most scenarios of global environmental damage such as nuclear winter, 5–10 times as much as is currently stored. Holding facilities would be expensive, as would pest control. In addition, there is the moral issue of storing food when there are people starving now. Almost all national food reserves have the goal of stabilizing market prices for crops that are seasonally available. We do not manage food reserves at the global level. A system to manage global food stores against global disaster would also be extraordinarily difficult to manage in event of such a disaster. Who would get the food and how? I could find no estimates of how much such efforts would cost overall, and even the strategy of stockpiling food needs substantially more research [140].

Costs for Protection Against Pandemic and Bioterrorism

The costs to prevent spread of novel diseases or newly evolved diseases are clearly less than the potential costs of diseases left unchecked. Once companies produce vaccines against a specific disease, governments and individuals who purchase the vaccines absorb the costs. This, unfortunately, leaves less well-off individuals and governments unprotected.

A secondary issue is how to support the capacity to produce a large quantity of vaccines rapidly. The demand for flu vaccine is relatively continuous, as is the demand for regular immunizations. In contrast, when the H1N1 (swine flu) pandemic was prevalent, the amount of vaccines required against flu immediately doubled. Similarly, an epidemic of yellow fever in Brazil in 2017 led to vaccination shortages. Creating large amounts of vaccine requires injection of infectious agents into chicken eggs, where the virus replicates, and then harvesting the virus from the eggs. Maintaining this capacity when there is not a continuous need for vaccines is very expensive and wasteful. While companies that produce vaccines are willing to set up facilities to make regular inoculations from which they profit, most companies not able to profit from sporadic outbreaks because maintaining such equipment and facilities is an ongoing cost.

We need to support a global surveillance network if we are to identify emergent diseases before they become widespread enough to explode globally. The World Health Organization coordinates the Global Influenza Surveillance and Response System, which is made up of 138 national centers funded by their own governments. In 2012–2013, under a reduced budget, the World Health Organization put $16 million into collaborations to manage emerging diseases. The United States joined a consortium of almost 30 other countries and the World Health Organization (WHO), the Food and Agriculture Organization (FAO), and the World Organization for Animal Health (OIE) to combat emerging diseases. The United States has put $40 million toward this effort.

Individual nations also put billions of dollars into disease research and control. In the United Sates, Health and Human Services spent about $6.15 billion on response to the H1N1 pandemic. These numbers do not lead to an exact estimate for long-term monitoring and preparation for a global pandemic, but the total to do this correctly would be well over $100 billion across the globe.

According to a US General Accounting Report on Bioterrorism in 2001, the United States alone put $156 million each year into agencies to prevent bioterrorism and terrorism. After a terrorist sent a series of letters containing anthrax to officials in the United States in October 2001, US biodefense spending rose to $4 billion for actions to detect and prevent bioterrorist attacks, which includes stockpiling vaccines.

Costs and Progress in Reducing Nuclear Weapons

The United Nations Institute for Disarmament Research released a report in 2003 detailing costs of disarmament [141]. The report outlines costs of programs to reduce the number of nuclear arms and balances them against the costs of maintaining nuclear arms. The report indicates that the United States accrued savings of $1.5 billion from 1991 to 2001 associated with the Strategic Arms Reduction Treaty between the United States and the former Soviet Union. The report cites the Department of Energy as estimating the cost to the United States of $30 billion for cleaning up the nuclear weapons complex. Costs of maintaining nuclear weapons are high, and costs of monitoring other nations for compliance to arms control treaties are also high.

Summary

While costs of solving many of these problems seem high, for the most part, they are not that great relative to the total scale of the global economy. It seems that the major impediments to solutions to these problems are not monetary but relating to human nature and how human nature shapes society and politics. Many of the efforts to solve problems will eventually pay for themselves. Controlling diseases means less loss of work productivity. Protecting the environment means goods and services that it provides to humanity will continue to be available. Educating and feeding people means increased economic productivity. The political difficulty lies in figuring out who will pay. An equitable global scheme for payment will be difficult to attain as each country would like others to bear the burden of the cost. Should the amount each country pays be based on the number of people in the country or the total income of that country? The root of this question lies in human behavior. Can we work cooperatively to solve the worst problems in the world? There are positive signs; the very reason humans are such a successful species is cooperation, and I discuss the science behind cooperation in the next chapter.

Chapter 11
Technochimp: An African Savannah Survivor Looking for Solutions in the Modern World

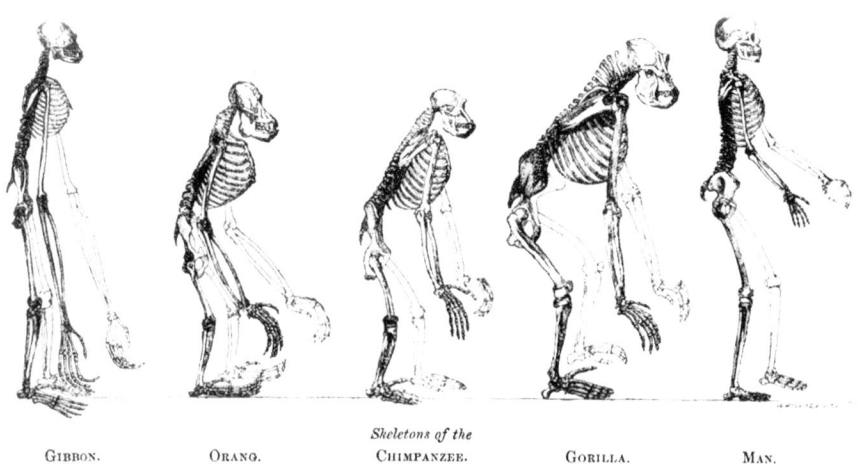

Skeletons of the

GIBBON.   ORANG.   CHIMPANZEE.   GORILLA.   MAN.

Photographically reduced from Diagrams of the natural size (except that of the Gibbon, which was twice as large as nature), drawn by Mr. Waterhouse Hawkins from specimens in the Museum of the Royal College of Surgeons.

Frontispiece to Huxley's Evidence as to Man's Place in Nature (1863)

For most of the evolutionary history of humans, we were a species that lived in small groups, hunting and gathering. All the while, we were avoiding predators and attempting to manipulate the environment to better suit our needs. Cultural evolution started slowly; we invented simple weapons and tools and acquired the ability to use fire and to clothe and shelter ourselves over the first millions of years that our ancestors inhabited Earth. Still the basic main concerns were always present. Will I eat and not be eaten today? Will I find a mate and protect my offspring? Can I make myself more comfortable?

We still have these basic concerns, but the context is radically different. Now instead of small family groups, we need to cooperate and interact with thousands of people over a lifetime, often tens to hundreds in any individual day, many of them we barely know. Our ability to alter the environment around us and do work is vastly greater than ever before. Only kings could command 1000 horses to do their work up until recently, and even kings with dominion over many people are

© Springer Nature Switzerland AG 2019
W. Dodds, *The World's Worst Problems*,
https://doi.org/10.1007/978-3-030-30410-2_11

very recent phenomena relative to our entire evolutionary history. Now I can call up a truck with a 1000 horsepower engine and have it work for me for a modest sum. People in the place where I live commonly drive vehicles that are 250 horsepower or more.

We communicate in milliseconds and move materials and ourselves around the world in days. Society has become extraordinarily complex with finely divided tasks and expertise. Our ability to adapt to new situations and cooperate with large groups of people has led us to amazing achievements, health and happiness for many, art and entertainment, and incredible intellectual progress. Still, we are our basic biology, and the evolutionary basis of our very being guides how we behave and make decisions.

We are almost genetically indistinguishable from the other apes. No behavior in humans has been identified that does not at least have a rudimentary equivalent in other apes [142]. The more we look into it, human behavior is heavily influenced by biology. Still, as a species, we have done pretty well for ourselves so far. Cultural evolution has allowed us to exceed our original primate origins and construct modern civilization. Now advanced research methods allow behaviorists to study the links between brain activity and behaviors such as altruism [143]. At the same time, social scientists are establishing the limits to predicting behavior [144].

Our Inability to See the Big Picture: It Never Mattered Before, Why Does It Now?

Many civilizations rose and fell. Well-developed civilizations, such as that on Easter Island, completely fell apart. Empires crumble, and others move in to take their place. The people either moved elsewhere, adapted, or did not survive. Contemporary society is global, and this is the first time humans have had the ability to influence the entire planet. We have no place else to go if the planet becomes uninhabitable due to our actions or outside factors. Each of the problems that I have discussed requires global cooperation to solve, yet our species has no history of solving global problems, only a history of failed civilizations. We have the hubris to think that this is the exception, that the first time we have a global civilization, it will not crumble. It is possible that this is true; this book is an attempt to consider what it will take to solve problems in spite of some of our engrained behaviors.

We have come to dominate the planet based on the ability to predict how our actions will influence our future survival. This has been accomplished in part by cultural evolution adapting to basic components of our genetic program to allow broader cooperation and specialization within societies. This general genetic program with its variants has come about from natural selection, and it is difficult to see how natural selection has shaped humans in such a way that they have an innate ability to think and act globally. This leaves the question, what do we have to work with?

Marlene Zuk [145] argues that a rigid view of humanity as "still on the savannah" is misguided. What she means is that we are not the same as the first humans that evolved on the African savannah. Part of her argument is in response to the "paleo diet," where people imagine how we can only be healthy eating like our Paleolithic (Early Stone Age) ancestors did. She is correct; the human species is still evolving and has adapted to agricultural diets. Some of us have evolved to consume milk into adulthood (retain the ability to digest lactose), populations have evolved to live at high elevations, and other genetic adaptations have arisen related to disease resistance and diet. However, evolution has hardwired the most fundamental aspects of our behavior (e.g., the urge to survive and reproduce), and these still drive many of our actions.

The fate of most species is extinction. Many species have gone extinct because evolution has driven them down the wrong path. I hope that humanity will evolve culturally or biologically before any of the worst problems cause our extinction.

Selfishness: Why Most People Don't Care if You Die

Most people on Earth do not care if you die. Most people care more about their own deaths than almost anything else. Our relationship with death and dying is fascinating to me as it gives us a glimpse into the deep evolutionary roots of our shared behavioral characteristics.

Almost all animal behavior is ultimately aimed at surviving, growing, and reproducing. This is the expected outcome of natural selection. Organisms with characteristics that allow them to survive, grow, and reproduce more successfully than others pass their genes and have more offspring. Animals with simple behavior have it all hardwired. Animals with more complex brains and behaviors not only have the hardwired behaviors that keep them alive but also have learned behaviors that help them be more successful.

If you trip and your head starts rushing toward the ground, there is a real danger that your skull will break and you will die. The physiological response is impressive, as the adrenaline makes the fall seem almost in slow motion and you put your arm out to keep that melon on your shoulders from smashing to bits on the pavement you are rapidly approaching. This is clearly very hardwired. Hunger, thirst, thermoregulation, and avoiding immediate danger are all deeply engrained evolutionary characteristics; most of us, most of the time, behave just as most all other animals on the planet do.

There are times when people worry less about dying. The most fearless people tend to be young males who are willing to fight and risk everything for potentially very little gain. Still, this makes evolutionary sense. One human male can fertilize many human females, so a shortage of males is not as bad as a shortage of females with respect to continuation of the species. Aggressive young males can get more resources, expand territory, and defend weaker members of the group. So why would they want to risk dying?

The idea of kin selection explains the roots of such behaviors. Take, for example, a young male in a small family group early in the evolutionary history of *Homo sapiens*. This male shares many genes with other members of the group. If he dies, most of his genes still keep going. If he dies in a way that insures more of his genes are passed through his relatives than would be if he survives, then there is potentially positive selection for the action that caused his death. If his death allows survival or success of the group relative to others, then one can imagine how evolution could select for the genes for behaving aggressively on behalf of the family group.

Cultural evolution has expanded and played upon the behaviors ingrained by kin selection as the groups got bigger and society becomes more than isolated groups of one or a few families. The "us-and-them" mentality is evident in nationalism, support of sports teams, and still in supporting and protecting immediate families.

Back to death. When a close family member dies, the response is not intellectual. It is as if you have been physically stunned. You are disoriented, extremely agitated, and perhaps angry about it. It is hard for me to argue that such a fundamental physical response is not a deeply evolutionary rooted behavior. Now consider when a distant relative or maybe somebody you didn't know well dies. You feel bad about it for a while and maybe shed some tears. The level of anguish and grief does not approach that which comes from loss of a close family member.

It makes evolutionary sense for a person to feel deep pain from the death of a family member. The family member shares many of your genes. If the death makes a big impression, maybe you can figure out a way to avoid related future things happening to yourself or others in your family. Protecting your family against death protects your genes. Maternal and paternal care are the strongest examples of this.

Around 150,000 people die on Earth each day according to the World Health Organization. Many of these deaths are natural, but many are tragic and avoidable (as I already talked about in this book). There is no way we can feel as strongly about each of these deaths as we would about the death of a family member or even a distant relative or friend. Furthermore, humans probably would not have developed civilization if they felt strongly that other people should not die. History of victorious civilizations is full of tales of slaughter of opponents.

An interesting study indicates that when death of others is viewed through the lens of an economic marketplace, there is even less regard for suffering of other species [146]. The willingness to support the care and upkeep of mice, or kill them, was less when governed by a market. In a first experiment, researchers offered individuals 10 Euros. If they took it, the experimenters would kill a mouse. If they did not, the researchers would house and feed the mouse. In a second market-based experiment, a seller was given 20 Euros, and a buyer negotiated with the seller. They could repeatedly bargain over the fate of the mouse. If the seller and buyer could agree on how to split the 20 Euros, then the researchers would kill the mouse. If they could not agree and forfeited the money, the mouse would live. The mouse lived 54% of the time in the first experiment. The bargaining market pairs saved the mouse only 28% of the time.

Of course, you could argue that many people die altruistically or against their own better interests. People die by suicide, and there is no good evolutionary reason that they should. Soldiers fight battles that will do them no good regardless of the outcome. Saints and suicide bombers risk or experience death for what they perceive as higher good. I argue that most people do not behave this way. If we are considering global patterns, what most people do is what matters.

I think most people behave selfishly, that this has deep evolutionary roots, and we should consider this if we are to solve global patterns. On the positive side, selfishness can also lead to cooperation, as we have a society where nobody can accomplish everything alone. The trick then is to understand when self-interest can lead to cooperation. The field of game theory explores just this issue, and there has been a recent huge explosion in research that helps us understand the ways people behave when confronted with a choice between cooperatively solving a problem and going it alone.

Game Theory and Predictability of Unpredictable Behavior

Several "games" have been set up to understand cooperation. One of the oldest behavioral research games is the "prisoners' dilemma." It goes roughly like this. Police arrest two criminals for a crime that they committed together and place them in separate holding cells with no way to communicate. If neither confesses, they both get a year in jail on a lesser charge. If only one confesses, he goes free, and the other gets 3 years in jail. If both confess, they both get 2 years in jail.

The prisoner who always confesses gets 0 or 2 years (1 year on average). The player who never confesses gets 1 or 3 years (2 years on average). Thus, the purely selfish approach would be to always confess. Yet when people play the game for one round, they tend not to play purely selfishly. If the players play the game repeatedly, then they tend to play more selfishly to punish the other player. This is the first indication that people go into relationships with some expectation of cooperation.

A second game is the ultimatum game: player 1 gets $100. They can offer any part of that to player 2, and if player 2 accepts the offer, they keep their respective shares. If player 2 rejects it, neither player gets anything. The players only play the game once, and the players cannot negotiate before the offer. Rationally, player 2 should accept any sum, because that is better than nothing. When the players play the game with real money, player 2 often rejects an offer. This game has been tested with 1762 players (881 pairs) across 15 societies, and overall, 56% of offers of $10 or less were rejected. The level at which offers were accepted varied by society, but globally there was a propensity to reject "unfair" offers at personal cost [147].

There are many variations of these games, and the general result is that people tend to cooperate when they view it as beneficial; they are willing to take a loss to punish those who are not cooperating. People do this calculation regularly, and it is possible to predict the overall results by assuming each player puts their own well-being first.

Such behavior is fundamental, as the games to test how people choose to cooperate yield similar results regardless of where in the world they are tested. At a basic level, they even apply to other primates, as chimpanzees view unequal rewards as unfair and the same part of their brain shows activity when they react negatively to unfair rewards, as in humans under similar conditions. Even if people learn the basics that dictate cooperation, they learn them early and in a similar fashion across cultures, so the behavior is predictable in general in a global sense.

Social norms are an additional key aspect of how humans cooperate, and this is much more fluid and culturally determined than the basic calculation of costs and benefits of cooperation. A social norm is an expected social behavior. There are very clear examples of how social norms alter human behaviors. One example is how attitudes toward smoking have changed in many areas of the world. As we enact more regulations on public indoor smoking, people have become more sensitive about how their smoking influences others. People alter behavior strongly based on social norms. In the worst case, people riot. They do things they would never do in isolation. On the opposite side, even noisy people tend to be quiet in churches and libraries. A social norm provides an idea of how people gauge cooperative behavior.

Education is one of the most important aspects of being certain that people react rationally and understand the consequences of their actions, as well as the potential punishments and rewards. If people do not believe a global problem exists (e.g., deniers of climate change and the fact that carbon dioxide released by burning fossil fuels is exacerbating the effect), then they have no incentive to alter their behavior (because they perceive no punishment for continuing the behavior). The other part of the equation is how people gauge punishment and reward once they have accurate information.

Punishments and Rewards

An essential aspect of punishment, reward, and cooperation is that people view cooperation as a good strategy more often when members of a group are interacting with each other and have the expectation to continue interacting in the future. That is, when people think of others as "us" rather than "them," they are more likely to cooperate. There are two reasons for this. First, closely related individuals of many species cooperate because there is an evolutionary advantage to helping an individual that shares a substantial number of your genes. This kin selection leads to behaviors such as siblings caring for younger siblings and parents favoring their young over others. Second, humans have been successful because they are the ultimate cooperative species, specializing in areas to receive goods and services from other areas. In this case, cooperation occurs because both parties expect to benefit. If either party reneges on the cooperation, then the other at least stops cooperating and may even punish the other party if it is possible. Cooperation takes a range of forms. The extreme example is family where individuals are likely to give the most to cooperation. In the intermediate case, people are more likely to help those in their own community than they help those in others. Nationalism is a broader extension of the cooperative idea, and global cooperation to solve problems is the broadest.

Modern societies are codified systems of cooperation with punishments for noncooperative behaviors. Laws codify how people cooperate (e.g., pay taxes to fund infrastructure, agree to refrain from robbing or murdering others). Systems of punishment enforce the cooperation, both formal and informal. In much of the modern world, punishment takes several forms. These include laws that fine or imprison people, social interactions that bring shame and ostracism (social norms), and taxes or fees that act to discourage specific behaviors (e.g., cigarette taxes may lower smoking rates).

Given the penalties and rewards, individuals then decide when cooperation is in their best interest and when it is more beneficial to cheat. Traffic laws set speed limits, but many people think the benefit of saved time is great enough that they are willing to break the law. They only do so because they view the chances of punishment as low. If they go much over the limit, the chance of getting caught and the amount of the fine goes up. They avoid the cooperation of making the road a safer place for all to drive because they place their personal benefits above those of the group. With any system of cooperation, some will cooperate, and others will cheat or defect with varied degrees of severity, because they weigh the costs and benefits differently.

Limiting punishments and rewards to a single aspect can distort cooperation to undesirable outcomes. For example, economic self-interest drives economic cooperation in the form of corporations. Such corporations can ultimately harm the system if they drive out all the competitors. Thus, we enact antitrust laws.

Using an economic "discounting" approach, the world's problems can lead to nonoptimal solutions if we discount other potential future benefits. However, there is no requirement to discount based solely on economic benefit. For example, the general concept of discounting plays a role in evolution, where natural selection favors organisms that maximize reproduction of future generations (pass their genes) and leads to behaviors that may harm individuals at the moment [148]. For example, reproduction is costly and potentially harmful, but all organisms need to propagate, or the species will go extinct. Cultural evolution could provide a similar solution to global problems. If humans can understand the consequences of their actions based on harm and rewards to future generations, some problems may be more easily solved (e.g., global climate change, the cost for asteroid detection, disaster preparedness).

The theory of cooperation becomes more intricate as social settings become more complex and with different ways to cooperate. One thing that destabilizes cooperation is second-order free riding. These are the people who cooperate but do not punish those who do not cooperate. Since punishing others requires a cost, the people who do not punish destabilize the system. A solution to this problem is to have all contribute resources to a punishment pool prior to the cooperation event [149]. One could view paying taxes and accepting society's laws as conforming to this approach. On the positive side, cooperation continues even when the penalty for not cooperating is lifted, probably because of establishment of a social norm [150].

People act directly to create punishments and rewards for other people, but with a global perspective, it is important to consider punishment and reward within societies and among different societies. It is important to understand how individuals perceive costs and benefits. If people do not think a problem is real, then they are very unlikely to try to solve it. Again, education and information allow people to accurately assess problems and make rational decisions on when to cooperate. However, some basic features of human nature work against this.

When Worldview Trumps Facts

You have a pain in your leg, and a doctor does some tests and tells you that it is a blood clot. She says you need to moderate your behavior and spend some money on medicine, or you could be much worse off, and even die. A reasonable person would seek a second opinion, and if that doctor also said the same thing, would do something about it. What if you really did not want to moderate your behavior? In that case, you would see another doctor. Consequently, you go to 100 doctors, and 95 of them agree that the clot is real, and you could die if you do nothing about it. Most reasonable people would stop at the second doctor. Almost nobody would deny the 95 doctors have a valid point.

Even though 95% of the experts link global warming and climate change to greenhouse gas emissions, many people around the world deny the linkage. Several concepts from social science explain why people react negatively to the clear warnings from scientific experts. The first general tendency is the degree to which people are susceptible to confirmation bias. Individuals tend to have a bias toward facts that confirm their worldview. The scientific approach revolves around methods to minimize confirmation bias. In contrast, many people only look for facts that confirm what they already believe. This is a familiar debate strategy but goes only so far in the real world where scientific laws do not change based on the wishes of people. This is why Al Gore referred to global warming and climate change as an "inconvenient truth"; confirmation bias allows people to dismiss inconvenient facts.

A second tendency explains the inclination of non-experts to ignore scientific facts and scientific experts to be unwilling to make definitive statements. The Dunning-Kruger effect is the general tendency of people who are less skilled or knowledgeable in an area to overestimate their abilities in that area and the tendency of experts to underestimate their abilities or degree of knowledge [151]. Thus, the public tends to think they know more than the experts do, and experts do not make their points as forcefully as they should.

With respect to climate change, there are numerous people (perhaps you know some) that have no training in climate science but have strong opinions. They believe that volcanoes explain increases in CO_2, the intensity of the sun explains warming, and that somehow the climate experts have ignored or worse, covered up, this obvious evidence. On the other side of the coin, scientists will always qualify answers, because they are never 100% certain statistically that they are correct.

Science has repeatedly gone through major changes in worldview because one incontrovertible fact leads to the overturn or modification of a scientific theory; in a purely scientific approach, you never can prove with absolute certainty something is true; you can only become more certain as science discounts each possible alternative explanation. Scientists are an argumentative bunch. They have the tendency to question everything and find the exceptions. This means that scientists are poorly suited to make strong impressions in the public arena, but people who are not experts and do not have a scientific approach have no qualms about saying they are certain their worldview is correct.

A third tendency, one that makes it even more difficult for scientific facts to direct public policy, is attitude or belief polarization [152]. Social scientists observe this polarization when they give some people reliable information (e.g., documentation of facts or expert analyses) that counters an attitude or belief. The information that contradicts their beliefs actually causes them to harden their attitude and intensify belief. The hardening of worldview in the face of contrary facts is also related to the belief persistence (people tend to be more likely to accept evidence that confirms their first idea about a topic) and the primacy effect (people weight early information more heavily than later information) [153].

As a scientist and an educator, I have often mistakenly thought that if you enter an argument armed with facts or teach students the facts, it will be enough to alter their beliefs. The social science indicates that this is often not true. Unfortunately, different "news" sources capitalize on these human behaviors and present falsehoods that confirm the bias of the people who consume that information. These features of human behavior explain why fake news on social media is transmitted so successfully.

The Individualistic Aspect of Cooperating to Solve the World's Worst Problems

Solving problems requires cooperation of the public. For example, eradication of a disease requires that all individuals be free of transmissible disease and often agreement that all should be vaccinated. Getting people to collaborate and governing cooperation is difficult and requires understanding the social science behind cooperation [154].

If the public makes solving the world's problems an issue, then the politicians will try to solve the problems. This will require individuals to cooperate politically. Creation of social norms and the transmission of the desires of the public into policy both can lead to solution of problems. As more countries become democratic, the more important it is for political leaders to please the public.

There is hope for individuals cooperating politically. For example, one of the key environmental dilemmas (and a dilemma around any limited resource that people exploit) is the tragedy of the commons. This idea was published in the early 1830s and posits that with a limited resource pool with no regulation, the winner is the one that uses the most resource the most quickly [155]. However, societies have found ways around this tragedy [156], demonstrating how cooperation can solve societal problems in the face of selfish behavior.

The game-theory studies I already discussed were done under artificial conditions to assess why people cooperate in the real world. Studies that are more realistic are required to make the social science on cooperation relevant. One survey in Ethiopia that focused on groups of people that had the responsibility for managing their areas of forest did just that. The groups would benefit in the future if the forest was maintained. Individuals in the groups could gain more benefit from overexploiting the forest (e.g., by harvesting and selling more firewood than their sustainable share), leading to overexploitation of the forest. Protection of the forest was more successful in areas where people put costly surveillance into action to enforce the group agreements [157].

Some factors work against general cooperation of the public, particularly when market-like exchanges are involved. Individuals are willing to make larger sacrifices to support moral values than markets are. In one example, the willingness to support the care and upkeep of mice or kill them was less when governed by a market. I described the experiment where the market increased the proportion of people willing to take money to allow the death of a mouse earlier in this chapter. According to this research [146], market-based solutions to moral problems (e.g., should we feed starving people?) could face a steep barrier to solution based on the way people interact.

Altruism, and thus cooperation, may be a fundamental part of human nature. The success of humans in past times depended heavily upon cooperation, and we could have genes for behaviors or culturally transmitted behavioral patterns that are still in the human population but based on what it took to survive in the past, but not today [158]. If this were true, then the trick to solving the worst problems lies in part in finding ways to tap these innate behaviors and not some that would be more destructive. Additionally, we need to translate local individual progress toward solutions to international cooperation.

Chapter 12
International Cooperation

The lion and the mouse, Illustration from a collection of fables, circa 1867 Three Hundred Aesop's Fables; Literally Translated from The Greek by the Rev. Geo. Fyler Townsend, M.A

Countries could behave even more selfishly than individuals do. It is not clear how strong the social norm is among countries. There has not been any expectation that countries should respect human rights within their own boundaries until recently, and many countries still do not in spite of condemnation by other countries. Countries behave deceitfully, aggressively, selfishly, and disrespectfully. However, at times countries do cooperate. Sometimes they do all of the above at once.

Countries cooperate on many things. If they did not, many of our modern conveniences would not be possible, and wars would be far more likely. War has become less common [159], and international cooperation and trade continue to increase.

These trends mirror general decreases in violence across most individual countries in the world. I assume that countries will cooperate mainly when it is in their own self-interests. Understanding why countries cooperate is important because most solutions to the world's worst problems require international cooperation. The game theory results for individuals that I describe in the previous chapter have relevance to international relations; the things countries do are the results of an aggregate of human behavior.

Why Do Countries Cooperate Sometimes and Not Others?

Many countries cooperate on global trade, communications, shared defense, economic recognition of monetary systems, allowing travel into and out of countries, in legal matters, and allowing individuals to live in different countries. These types of cooperation do not refer to a central authority, necessarily, but can be coordinated behaviors among countries.

International organizations that influence politics and behavior come in various forms and ultimately have to act locally. These include national governments, regional to international institutions (e.g., North Atlantic Treaty Organization, the World Trade Organization), international political groups (e.g., Amnesty International, Greenpeace), multinational corporations, and various branches of the United Nations (International Monetary Fund, the World Bank, the World Health Organization, the Food and Agriculture Organization). These bodies can act to solve or exacerbate the world's worst problems [160].

Countries cooperate when the benefits outweigh the costs. For example, international communication comes at the cost of maintaining infrastructure and making systems compatible. The economic benefits are much greater than the costs, as most of the costs are borne by individuals (e.g., paying for phone and Internet).

One could argue that countries would mainly cooperate when they had a shared crisis. Jared Diamond has considered when countries act to respond to upheaval [161]. He hypothesizes 12 characteristics that relate to outcomes of national crises. These include (1) consensus that the nation faces a crisis, (2) acceptance of national responsibility for action, (3) clearly delineating the problem that needs to be solved, (4) getting assistance from other nations, (5) using other nations as models for solutions, (6) national identity, (7) honest self-appraisal, (8) historical experience of past crises, (9) dealing with national failure, (10) flexibility, (11) national core values, and (12) freedom from geopolitical constraints. He assesses these 12 areas with respect to crises faced by 6 nations and finds support for the ideas. He also assesses these characteristics with respect to global problems and finds it difficult to apply all of them. International cooperation requires some of these characteristics (recognizing there is a crisis, accepting responsibility, delineating the problem, international identity, honest self-appraisal, flexibility, and global human core values). Others do not apply (getting assistance from other planets or using them as models, past history, and freedom from geopolitical constraints). There are obvious additional characteristics that do lead to international cooperation though.

One of the best examples of global cooperation to fix a problem is the agreement to limit chlorofluorocarbons (CFCs) to manage damage to the upper atmospheric ozone layer as I discussed in Chap. 7 on the global environment. In this case, the cost was mostly a modest economic impact related to finding substitute refrigerants and propellants, phasing out the CFCs, and phasing in the alternatives. The alternative potential cost was much greater, total devastation of civilization. If the ozone layer disappeared, most life on Earth would as well. The benefit of continued existence is far greater than the relatively modest cost of limiting the chlorofluorocarbons.

The lack of obvious effort of most countries on Earth to control greenhouse gas emissions or nuclear weapons is in stark contrast to the ozone control measures. The cost of full-out nuclear war is complete extinction of humanity. The perceived benefit is immunity to invasion by a country with more nuclear weapons. In this case, so far, the dominant nuclear powers (the USA and Russia) have decided not to destroy enough of their nuclear arsenals to guarantee there will never be total annihilation. The USA has taken another step backward and decided to pull out of a key nuclear agreement as of 2018, the Intermediate-Range Nuclear Forces Treaty, with Russia following suit.

The potential costs of global warming are tremendous in the future. If runaway greenhouse heating occurs, large areas of the Earth could become uninhabitable, including many coastal areas. The benefit of ensuring that future generations have an inhabitable planet is, so far, not great enough that the dominant releasers of greenhouse gasses (the USA, China, and India) have made any serious effort to control emissions.

War provides a good example of cost/benefit analysis. A country will generally only go to war if the perceived benefit (e.g., domination over another country, protection from being attacked by another country) exceeds the costs (economic, political, and human life and health). In this case, older men usually feel the cost of losing the lives of many younger men is not as great as the potential benefit of war.

One approach to viewing international interactions has been to assume that countries will be very unlikely to cooperate if such cooperation makes other countries stronger. Stronger competing countries could ultimately take over the weaker countries. This view ignores the established human propensity to punish noncooperators, leading to cooperative behavior coerced by the potential of punishment.

Attempts to apply game theory to international cooperation have been met with controversy among political scientists. An important question is whether decentralized regulation and enforcement leads to cooperation. Some argue that problems associated with political systems reward those that win, even if winning leaves people less well off than cooperation [162].

In "The Plundered Planet: Why We Must – and How We Can – Manage Nature for Global Prosperity." Dr. Paul Collier suggests restoring human relationships to natural capital requires international cooperation [163]. He suggests (1) education of people in developing countries is necessary to create a population that understands how richer people and corporations are plundering their natural capital, (2) society should not tolerate it when developed and developing countries are complicit in plundering the natural capital of poor countries, and (3) coordinating international actions is required for real progress to be made.

A Global Society

The idea is new that the world is a better place when people in all countries have basic human rights including adequate food, shelter, freedom from persecution for beliefs, and a say in how they are governed. Not all follow this, but it is becoming more prevalent. Steps toward this ideal have led to impressive gains in human well-being [164].

The Arab Spring exemplifies how people can become intolerant to repressive regimes as information and education allow them to see that there is another way. This also exemplifies what happens when a large segment of the population of a country becomes dissatisfied; revolution can occur, but the result is not necessarily better than what was thrown out.

Society balances the desires of individuals against the group as held in check by systems of governance. While we generally think of national governments being the key agents of governance, there are many other aspects. For example, a social norm in society probably controls basic behaviors more than specific laws. People in the USA have a norm of speeding by about 5–10 miles per hour on highways, regardless of the law. There is no specific law against picking your nose in public in many countries, but most people do not do that.

Successful international cooperation is certainly possible. Most countries readily exchange money, and it is possible to contact people through mail, phone, or Internet connections in most parts of the world. The global financial and communications networks are operational because people benefit more from cooperating than they lose by being part of the network. The benefits exceed the infrastructure costs, so people, corporations, or governments are willing to support the networks. However, the question remains, how well will we collaborate to solve global problems?

International Agreements, Agencies, and Organizations

Different problems require solutions by different agencies and agreements. Two of the largest problems, death and suffering from disease or hunger, are closely interrelated, and I will treat them together here. At a fundamental level, we can solve both problems by increasing the standard of living for the poorest people. We currently have enough food to feed the world, but distribution and autonomy to raise food are the major hurdles to decreasing hunger. Thus, the solutions to this problem are mainly political. The United Nations World Food Program is the largest agency fighting against hunger. As the average per capita income has risen, the proportion of hungry people in the world has dropped since the 1960s. Much of the progress has been in Southeast Asia; as China, India, and other countries in the region have developed their economies, they have decreased rates of poverty and disease.

The World Bank in 2000 lent $1.3 billion to improve agriculture, down by more than half from 1990. The data suggest individual countries need to solve world hunger based on local economic development. International markets are working against these trends as large corporations buy productive land to produce commodities preventing lower-income people from having the possibility of growing their own food. Some private philanthropic organizations have been helping reduce hunger and poverty. For example, the Bill and Melinda Gates Foundation provided the World Food Program $66 million for a program to increase small farmer income and $43 million for Heifer International, a nonprofit charity that helps small farmers increase dairy production.

International governmental agencies such as the World Food Program help raise awareness of the issue and coordinate efforts when famine occurs, and disaster relief is one area where international agencies and cooperation are needed. Both natural disasters and political instability can threaten food security, and in those cases, simply helping small local farmers is not enough to avoid short-term widespread hunger.

One of the most successful international agreements on environmental issues, or any issue for that matter, was The Montreal Protocol on Substances that Deplete the Ozone Layer that was the result of the Vienna Convention for the Protection of the Ozone Layer. Most countries have signed this agreement, making the Montreal Protocol and Convention document the first globally signed treaty in the United Nations. A key feature of the protocol is the provision that provides developing countries with funds to help them comply with the protocol. Thus, the agreement and subsequent control of release of ozone-destroying compounds were very successful, considerably more successful than other global environmental treaties. Unfortunately, China has violated the protocol [165], so continuous enforcement is necessary for this and other treaties to protect the environment [166].

The Kyoto Protocol to the United Nations Framework Convention on Climate Change (UNFCCC) set limits to greenhouse gas emissions with the goal of decreasing emissions to the 1990 level or somewhat lower by 2012 and eventually bringing down emissions to levels that will not disrupt the global climate. As greenhouse gasses have continued to increase globally since 1990, at greater rates than would be expected if emissions dropped to 1990 levels, it is obvious that the Kyoto Protocol has not been effective. The economic reasons for this are related to unequal pricing of greenhouse emissions [167].

So why are some major international agreements successful and others not? In the case of the control of CFCs to protect the ozone in the upper atmosphere, several important factors came into play: (1) The consequences of not controlling the CFCs could easily lead to such high UV levels that producing crops would become impossible, so the stakes for not acting were high. (2) There were readily available substitutes for CFCs, or their specific uses were not very important (e.g., propellant for spray cans). (3) The industry that profited from CFCs was not huge, so had modest political clout, and could make money producing or handling the substitutes.

In contrast, control of greenhouse gasses faces major hurdles: (1) A very wealthy group of international corporations with strong political influence has much to lose if fossil fuel use decreases. (2) Developing countries need fossil fuels to develop their economies. (3) The dangers of global warming are not immediate (although they are becoming evident), so it can be politically difficult to justify short-term costs for long-term gains.

As individual countries build nuclear weapons, international agreements among countries are necessary to control nuclear weapons. Political organizations, such as the Union of Concerned Scientists, can drive the resolve to make and hold to such agreements by organizing scientists globally to accurately portray the risks of nuclear war. Notable agreements to control nuclear weapons include the Partial Test Ban Treaty (1963; to limit atmospheric testing), the Nuclear Non-Proliferation Treaty (1970; to allow peaceful uses of nuclear energy and discourage development of nuclear weapons), Strategic Arms Limitation Treaties (SALT I and SALT II; to halt construction of new nuclear weapons), Anti-Ballistic Missile Treaty (to limit antiballistic missiles, the USA withdrew in 2002), and the Strategic Arms Reduction Treaty (ultimately led to removal of about 80% of nuclear weapons in existence, renewed in 2011).

The International Atomic Energy Agency is independent of the United Nations but reports to the UN General Assembly and Security Council. The agency promotes nuclear safety by helping avoid nuclear accidents in power plants and preventing nuclear proliferation. The agency relies upon cooperation of individual countries to act and is called upon to verify if countries are capable of producing nuclear weapons. For example, Iran claims it is not trying to develop nuclear weapons. It is necessary to validate those claims, and the International Atomic Energy Agency has attempted to do so.

The Bulletin of Atomic Scientists published its first Doomsday Clock in 1947, where the number of minutes until midnight signifies the degree of nuclear insecurity on Earth. It was set at 7 minutes to midnight when first published. In 1949, the clock moved to 3 minutes after President Truman told the US public the USSR tested its first nuclear device. In 1953, it moved to 2 minutes when the USA and USSR both tested the first hydrogen bombs. In 1963, the clock moved to 12 minutes when the USA and USSR signed the Partial Test Ban Treaty, ending aboveground nuclear bomb testing. The clock got as far from midnight as 17 minutes in 1991 with the Cold War over and the USA and Russia making large cuts in their nuclear arsenals. The clock has continued to move toward midnight since that point as the USA has withdrawn from various treaties. Also, the major powers have not rejected the strategy of rapid response leading to a strong possibility of massive retaliation for an accidental missile release [168]. Finally, the threat of terrorist nuclear attack has increased, and India and Pakistan stage nuclear weapons tests. In 2019 the clock is at 11:58.

Controlling disease is an area where international cooperation is clearly required. Starting in the 1850s, a series of International Sanitary Conferences convened to negotiate measures to stop the spread of cholera, plague, and yellow fever across international boundaries. By the 1950s, the World Health Organization had adopted international health regulations. In a world with rapid global trade and travel, cooperation is necessary to stop the spread of emergent diseases as well as diseases from areas where they are chronic problems. These agreements require intentional organization as well as the ability to create disincentives for nations that do not practice safe trade and travel [169]. The disincentives could include trade embargos and other economic sanctions.

Some of the biggest causes of death and suffering are diseases that are easily cured or prevented. Control of these diseases is closely related to efforts to reduce extreme poverty, as discussed above. On the bright side, there have been some fantastic gains in control over communicable diseases, and many people are less likely to have chronic disease.

Small pox was eradicated by the mid-1970s, although vaccines had been available since the early 1800s. This triumph was coordinated by the international community, assisted by the fact that humans are the only host for smallpox. Thus, infected symptomatic people are the only carriers. Identification of cases and subsequent quarantine of people who had contact with the infected individual coupled with immunization of all people could locally control the disease. Between 1967 and 1979, about $300 million were spent on eradication. Economic analyses suggest this is far less than the cost of vaccination, treatment, and lost productivity to the disease if it were not eradicated.

The triumph over smallpox could be repeated with polio. Private and public sources fund the Global Polio Eradication Initiative (GPEI). Funding comes from many nations, the World Bank, the Bill and Melinda Gates Foundation, and the Red Cross/Red Crescent organizations. The GPEI estimates about $5.5 billion will be required to complete eradication, but the effort is complicated by the fact that many cases occur in areas that are currently politically unstable including portions of Afghanistan, Nigeria, and Pakistan.

International cooperation is also required to control bioterrorism. The Biological and Toxin Weapons Convention is a treaty signed by 165 nations which bans developing, producing, stockpiling, and transferring biological agents except for defensive research. This leaves about 30 nations which have not signed the treaty. In addition, the international cooperation on control of bioterrorism is mostly voluntary, and treaties are not in place. Coordination of biological research, information transfer, and reagent sales is desperately needed but could probably be accomplished by a nongovernmental organization or a UN-mediated group. I am not aware of broad agreements to control bioterrorism at this time.

Avoiding catastrophic meteor collisions is a unique problem, in that the initial identification of objects could be accomplished by nongovernmental organizations. One attempt has been made at this already; the B612 organization was planning a Sentinel Mission to place an observatory satellite that would catalog most objects in the solar system large enough to be planet killers. They attempted to raise $450 million to fund the project and were not successful. Now they are exploring cheaper alternatives.

The US Congress in 2003 directed NASA to identify 90% of near-Earth objects greater than 140 m by 2020. NASA initiated several projects and to date has identified most objects large enough to destroy a substantial portion of life on Earth. Continuous surveillance is necessary to detect uncatalogued comets, and space agencies as well as the public are involved in these efforts. The International Astronomical Search Collaboration brings the public and smaller observatories throughout the world into the process of tracking objects on space images provided to volunteers online.

In general, global governance is up against some major problems. According to Jim Whitman from the University of Bradford, there are several reasons to be pessimistic about the ability of global governance to solve our problems [160]. First, our capacity to produce unwanted and dangerous conditions at the global scale is developing much more quickly than are global control mechanisms. Second, the network of global relationships is becoming ever more complex and difficult to govern. Third, the current architecture of global governance is an expression of power and special interests. He states that we need to conceptualize global governance, which is in large part the topic of the next chapter.

Chapter 13
Consilience, Global Socioeconomic Political Enlightenment, and Socioenvironmental Restoration

National emblem of Greece (the Phoenix) from the Newspaper of the Government (1832)

Global cooperation is without a doubt required to solve global problems. Achieving this cooperation will require not one solution but input from many different segments of society and many different disciplines. The scientist and popular author E. O. Wilson pulled the old term *consilience* into the modern age to argue that solution of global environmental problems would require a bringing together of many disparate fields [170]. While we can consider the international or political hurdles to solving global problems, ultimately the ability to solve problems comes down to how humanity reconciles the fact that many of our individual urges contradict the requirements for long-term survival and happiness of global society as a whole.

A focused example of multidisciplinary collaboration for solving problems effectively was documented by heart failure researchers [171]. They found that heart

© Springer Nature Switzerland AG 2019
W. Dodds, *The World's Worst Problems*,
https://doi.org/10.1007/978-3-030-30410-2_13

failure patients were less likely to be readmitted into the hospital with multidisciplinary approaches rather than the traditional approach of simply receiving instructions from a doctor. In this case multidisciplinary approaches meant that post-release treatment included patient and family counseling, dietary assessment, nursing and social-service interventions, support groups, and specific measures to improve adherence to post-release instructions. Even a problem that the medical community has worked on for years is solved more effectively by broadening cooperative aspects of care. Here I discuss how solving the world's biggest problems requires changes in economies, politics, and society. It will also require multidisciplinary approaches to restoration of a sustainable relationship between society and the environment.

Consilience

Spectacular failures can occur when people do not consider knowledge across disciplines, and social change can come from many areas. Solving global problems will require social scientists, economists, politicians, policy makers, engineers, biological and physical scientists, artists, educators, and religious leaders. For example, historical analysis combining social sciences and humanities has successfully led to analyses that predict structure of human societies [172]. Human behaviors and physical realities drive global problems, so solutions absolutely require physical and social sciences to point out problems as well as ascertain feasible solutions. Once we identify the problems and causes, humanity's next step is to find solutions. These solutions could come from science, engineers, and social scientists. Implementation of widely unpopular solutions is not likely. This is where artists, educators, religious leaders, and others can shape public opinion and move policy makers, politicians, and other leaders to actions.

Likewise, large-scale environmental problems involving shared resources require multiple levels of coordination to avoid overexploitation. Attaining global sustainability requires systematic approaches considering coupled human and natural systems, socioeconomics, and understanding and responding to complexity and complex interactions [173]. The successful approaches to avoidance of problems include multiple layers of government, flexible management approaches, feedback of all involved groups, and proven methods of enforcement. Similarly, systems engineering seeks to find solutions to problems by taking a holistic view and considering many aspects of the problem that require different specialization, such as considering the full lifecycle of the project, reliability, risk management, and logistics. Newly emerging specialties will also be important; systems scientists (whose specialty includes general study of how complex systems interact) are learning to control complex networks [174] and applying methodology such as neural networks, machine learning [175], and artificial intelligence.

So far, I have discussed the role of sciences in identifying and solving problems and what actions would be required as well as some estimates of cost. Additionally, I covered the topic of how social science informs us of the fundamental aspects of human nature. This leaves open the question: how can we actualize change?

A Pew Research Center survey of people from 2010 in more than 230 countries [176] found that "Worldwide, more than eight-in-ten people identify with a religious group" and that "There are 5.8 billion religiously affiliated adults and children around the globe, representing 84 percent of the 2010 world population of 6.9 billion." These data make it clear that religion is an essential component of the lives of most people and plays a large part in how they view right and wrong.

There certainly has been debate about the role of religion in solving the world's problems. John Lennon imagines a utopian vision where there is nothing to kill or die for, including religion. Sectarian violence causes many world problems, with recent increases in the incidence of hostility related to religion globally. While such conflict has been common now and in the past, religion must play a part in solving the world's worst problems.

Many world religions directly address some of the world's problems (e.g., helping the poor and the hungry), but other issues are less directly addressed. There is no historical precedence for how religion would deal with global problems such as pandemics, nuclear war, and global climate change. Still, if religion promotes obligation to society, particularly for all people, then this could help solve some of the world's biggest problems. Many religious organizations dedicate themselves to making the world a better place and have activities directed toward this goal.

Artists, writers, athletes, and musicians provide much of the entertainment that we experience today and could be tremendously influential in helping solve the world's problems. They can be particularly important in swaying public opinion. Now that social media can spread ideas so quickly and cheaply, and getting social media noticed requires skillful presentation of issues, the potential role for the humanities in solving the world's problems is greater than ever.

Any of the human-related problems on Earth involve, in some way, economics. Some researchers in economics are concerned with global environmental problems (e.g., ecological economics) and others on how to reduce poverty. Amartya Sen received the Nobel Prize for his studies on welfare economics. In 2009, Elinor Ostrom won the Nobel Prize for her work on cooperation and common-pool resources (the commons).

A New Model of Relationships Among Global Society and Economics

Humanity will require global socioenvironmental restoration to allow sustainable habitation of Earth with respect to the ability of the planet to support ecosystem services (e.g., clean water and air, good food, stable climate) and to allow other organisms' continued existence. For example, the current model of burning fossil fuels as our primary energy source is clearly not a sustainable approach for local air quality and maintenance of global climate within ranges conducive to supporting humanity, yet large segments of our economy rely upon this one energy source.

Economic analysis suggests that the current pattern of putting the economic costs of human activities onto others is pervasive and leads to unsustainability [177]. Economists in part avoid considering economic costs by simply ignoring them as "externalities," but solving this issue is central for solving environmental and other problems [178]. For example, a forester gains economically from logging a forest. Once he cuts the forest, erosion rates increase causing damage to fisheries and water quality downstream. Under most current models, the forester is not required to pay the costs incurred on others. Socioenvironmental restoration will be required to change this type of interaction. Some argue that developing countries will become sustainable once they have developed a higher standard of living. However, even though developed countries have different socioenvironmental structures, neither developed nor developing countries are sustainable at present [179], suggesting a new paradigm of socioenvironmental relationship is necessary.

People tend to use resources without regard for future generations. Many of the problems I discuss (e.g., food security, environmental protection) require consideration of future generations. Social science research shows that games of resource exploitation almost never conserve resources for the future. This is because even if most people conserve for the future, a few defectors can use the resource above sustainable rates. However, if resource use rates are decided by voting, defectors are held in check by the majority [180]. Thus, democracy offers a route to solution of problems into the future. Democracy has increased worldwide since the 1970s [181] but has been threatened by the rise of populism and movement toward more authoritarian political figures globally [182].

One of the key issues in attempting to ease death and suffering is economic inequality. The vast gulf between the haves and have-nots in each country, as well as the economic disparities among countries mean that people on the bottom rungs of the economic ladder suffer disproportionately and needlessly. The fact that we have enough food to feed everyone on Earth, yet people are undernourished even in developed countries, and many of them in developing countries, is clear evidence for the role of economic disparity in global death and suffering. An analysis of economic disparity by Marten Scheffer, an applied mathematician, indicates that extreme economic inequality is inevitable in a global economy [183]. This suggests global-scale wealth equalizing institutions would be required to mitigate this inequality. Global cooperation would be required for this mitigation, but unfortunately, the powerful people globally are those with the most to lose from increasing economic equity.

Jeremy Rifkin, a key advisor to the European Union with respect to energy strategy, has laid out one potential route to a more sustainable energy approach. He proposes five pillars for what he envisions is the Third Industrial Revolution [184]: (1) move to using mostly renewable energy, (2) shift to smaller-scale power generation (e.g., wind, solar, biomass), (3) develop ways for smaller-scale entities to store energy (such as hydrogen), (4) integrate energy information so that small-scale energy producers can move their energy across the grid to others that are currently in energy deficit (using information sharing through the Internet), and (5) moving toward electric vehicles that can use the energy grid.

This transition to energy sustainability would clearly require large changes in economics (dominance of energy companies), infrastructure (constructing small-scale power generation, storage, and a more efficient energy grid), and cultural shifts. This would be a difficult transition, and certain segments of society will resist it. All of the aspects of social change (education, changing social norms, changing views of aesthetics) will need to come to bear to accomplish this transition. Parts of this transition, in particular movement toward alternative energy sources and reliance on a distributed network, are already developing rapidly.

The benefits to transition to more sustainable energy would entail more than just decreases in emission of greenhouse gas. Developing inexpensive technology for local energy generation will make the lives of people in developed countries better and allow them to become part of the global community. I travelled in Mongolia and visited "black markets" where thriving businesses sold solar cells, batteries, and power control equipment for their yurts (gers) in remote areas. Having electrical power in remote areas allows for communication (cell phones) and receiving news from the outside. Medical emergencies are much easier to address with outside help. These are all things that make life easier and safer, as well as more enjoyable. This technology and access allows improvement in the standard of living while maintaining the traditional nomadic pastoral lifestyle in tandem with the benefits of modern technology.

The "new" model would require several aspects of governance to be successful. Common resource pools are vulnerable to overexploitation because there is no incentive to conserve if others are using the resource and they do not agree with the importance of maintaining the resource sustainably. The scales of global problems require special approaches. Strategies that help solve resource use problems at this scale include (1) dialog among stakeholders and those that can provide information, (2) complex and layered institutions with redundant functions, and (3) ability to adapt decisions to current and changing conditions [185].

People, Thresholds, the Internet, and Intangible Predictions

Solving the worst problems requires money and political power. A number of the problems listed here are minor concerns to powerful people with money. For example, how many people that are wealthy suffer from hunger? Still, many problems relate to the well-being of all (e.g., avoiding global nuclear war, climate change). How is it that people work against their long-term interests? Why do they bite the hand that feeds them? Maybe it is inertia, maybe short-term greed, maybe a feeling of powerlessness. However, one way or another, we move forward in ways that benefit humanity. What will be required to redirect society such that it addresses and solves its biggest problems? How do we predict if social change will be positive?

Solving the problems will require a fundamental change in core values as identified by Jared Diamond as one of the 12 factors relating to crisis outcomes discussed in the previous chapter. By this I mean that people of the world will need to mostly adopt and act on the moral assumption that I laid out in the second chapter; death or suffering of any person, once they are born, now or in the future, is weighted equally across the world. The specific nationality, race, class, gender, sexual orientation, or religion of a person does not mean that they deserve death or suffering any more or less than any other. Successful adoption of this moral view would mean developing a "global identity." This does not mean that people need to act against their best interests, but rather that their best interests are served when global problems are solved.

Solving the problems covered here will require large-scale social change. One positive aspect is the explosion in social science helping us understand how to enact social change. For example, social tipping points occur when around 20% of people in a social network agree that social conventions should change [186] so we do not need to convince everyone that the worst problems need to be solved. Social norms can change drastically (e.g., smoking in public, foot-binding in China, littering in the USA). It is not yet clear how social norms could be tipped into a new state that will solve rather than exacerbate global problems [187]. It is also worrisome that society could transition to a state that exacerbates many of the social problems because those driven by short-term gain have the same information on social science and general public behavior. Witness interference in the 2016 United States elections and many in Europe via social media.

Social science research suggests the possibility of changing social views that lead to problems (or sway them for other purposes for that matter). For example, research on social contagion on how behaviors spread using a sample of 1.3 million Facebook users shows younger users are more susceptible to influence, men are more influential than women, and women influence men more than they sway other women. In addition, persuasive individuals tend to cluster in networks, so manipulating people with influential friends would spread behaviors most quickly [188].

Social science research also indicates that diversity helps us make good decisions. A study of collective intelligence of groups versus individuals found that groups are indeed more intelligent than isolated individuals. That study found that social sensitivity of group members was more important than intelligence of individuals in the group. They also found that a greater proportion of women led to greater collective intelligence [189]. This is one study of many indicating that diverse groups are superior problem solvers.

One of the problems people have is making predictions outside of conditions that they have experienced previously. Scientists are particularly plagued by this problem as they use prior observations to make formal predictions about system properties. Sometimes a system undergoes radical change and takes a long time, if ever, to get back to the same place. Scientists call this rapid change a threshold. The new condition is a stable alternative state. Social thresholds and environmental thresholds are both relevant to the materials I consider in this book.

Those wanting to destabilize other countries (e.g., Russian interference in global elections) likely are basing their approaches on social science. People that want to alter our behavior via media (e.g., advertisers, particularly political advertisers) probably also base their approaches on social science research. There is no reason to not use the same research for positive outcomes or at the very least to stave off harmful influences.

Understanding environmental thresholds is also required to avoid environmental meltdown. One of the most concerning examples is a runaway greenhouse effect where we cross a threshold that causes greater release of greenhouse gasses and leads the Earth to cook itself. This could occur because of permafrost melting or release of large stores of methane under the ocean. Another concerning issue is the destruction of biodiversity. Once a species is extinct, there is no way to bring it back.

The runaway greenhouse effect is one example that shows the difficulty in making solid predictions about thresholds. Scientists are careful to say they do not know if or when we will cross the threshold leading to runaway warming. They do understand that such thresholds exist and there is a finite danger that we will push past the point of no return (or already have).

We could push across a societal threshold and experience communal collapse in large regions of the world following a disaster, exacerbating many of the problems I have already discussed. A large volcanic explosion, a limited nuclear war, or a large asteroid collision could all come with a threat of global cooling and failure of crops. Failure of crops could lead to rioting and unraveling of society, not to mention widespread starvation. Civil preparedness such as food stocks, evacuation procedures, backup plans to feed people in need, and ways to provide clean water can help minimize the possibility of societal collapse that exacerbates death and suffering following a global-scale disaster.

A huge unknown is what role global connectivity and information sharing will play. Only recently have the majority of people on Earth been able connect with each other. Facebook is an example. This social network launched in 2004 but really took off by 2008. In 2018, there were 2.23 billion users, about 30% of the Earth's population. Facebook was a major catalyst for the Arab Spring, and Russia and other countries are using it as a political weapon to influence global politics. Many other countries and institutions are attempting to manipulate it to their own purposes, be they economic gain or political power.

We now live in a world where information spreads very rapidly, bad information spreads more rapidly than good information, and some information can lead to substantial social changes. Facebook, Google, and other companies collect our information, and it is used to produce detailed and specific advertising to convince people to buy products. However, personal information can be used to serve political and social purposes as well. There are open questions: can humanity harness the Internet and information for good, or will it be the agent of catastrophic social change? How can we forecast the future in the face of so much information?

Social science is finding ways to make better predictions. One route is to rely on "superforecasters" [190]. This type of person was recently described based on a series of forecasting tournaments constructed by the US intelligence community to create probability estimates of geopolitical events. The questions were about real future events such as: Who will be the president of Russia in 2012? Will North Korea detonate another nuclear weapon in the next 3 months? How many refugees will flee Syria next year? Participants submitted probability estimates over a

9-month period, and the individuals that were correct most often were identified after the events actually occurred. They then created teams from the best forecasters and compared them to regular forecaster teams. The superforecaster teams consistently outperformed the other teams. The researchers identified several characteristics that were consistent hallmarks of superforecasters' performance. The best worked hard (tried to forecast many questions), updated their beliefs more often than poorer performers, engaged more frequently with their teams, gathered more data, and were more willing to question the other group members. This study suggests that we could form groups of experts to help understand global problems, given attention to the makeup of the groups. Stated differently, a well-constructed group of experts will provide better predictive ability than a random group of individuals.

Enlightenment, Hope, and Fixing the Worst Problems

Hope springs eternal. Hope is the thing with feathers, Phoenix, arising from the ashes. Why bother reading this book or writing it if there is no hope? Hope is what gives humans the hubris to think that we are the one species that will not go extinct even if all others have. Hope is what allowed the slaves, those in the concentration camps, the people under grinding poverty, the victims of intended genocide, the lonely, the hungry, the hurt, and the abused to retain sanity and maybe even survive. Ben Franklin is reported to have said, "we must all hang together or most assuredly we will all hang separately" after signing the Declaration of Independence. I hope the world can work together instead of each country going its own way and ignoring the fact that selfish best interest in the short term is not the best for most in the long term.

Can we fix the problems? Some yes, others no. Protecting against some of the worst disasters is impossible. A direct hit from a gamma-ray burst would extinguish all life instantaneously. We have ways to hedge our bets against many aspects of these problems. A modest number of nuclear warhead detonations, an off-axis gamma ray burst, a mega "ish" volcanic eruption, a degree of societal breakdown, a disease that kills many but not most people, and many other disasters can be prepared for. People tend to favor the status quo, even though they don't necessarily believe that is the case [191], so change will be difficult.

The degree of poverty, starvation, and disease in the world is one of the most easily solved problems, though factors conspire to prevent solving it. People treat others terribly for their own advantage, and this will never change. Still, we are feeding, housing, and protecting an ever-greater proportion of the world's population. People, in general, are doing better. Recent history gives us hope.

I can envision a future with minimal human poverty and few people with preventable diseases. I can image social justice for all and limited probability of catastrophic nuclear war. It is not out of our reach. We are capable of great things and capable of including all people in these things. We should all hope and strive to make our dreams of a better world become reality.

Correction to: By the Numbers: Ranking the Problem

Correction to:
Chapter 8 in: W. Dodds, *The World's Worst Problems*,
https://doi.org/10.1007/978-3-030-30410-2_8

The original version of this chapter was inadvertently published with incorrect version of Figure 8.2. The correct figure has been updated in the chapter.

The updated online version of this chapter can be found at
https://doi.org/10.1007/978-3-030-30410-2_8

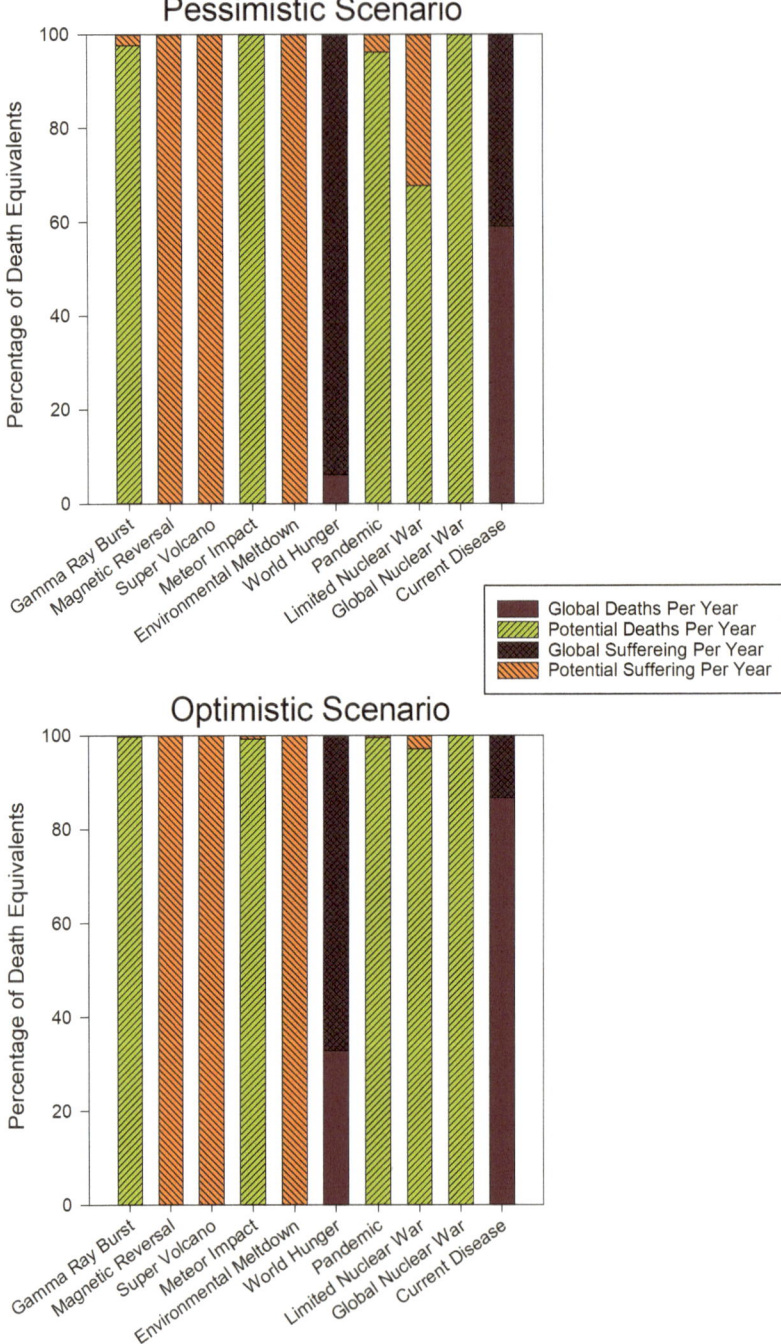

Fig. 8.2 Percentage contributions of deaths now, suffering now, future deaths, and future suffering to pessimistic (worst case and 20 years suffering = one death) and optimistic (best case and 150 years suffering = one death) scenario to the overall index

Appendix I: Calculations of Problem Severity

My approach to calculation of problem severity allows assessments of relative risks and playing them off against each other. First, I will define my terms and then put them into a single index of problem severity. Then I will explore different outcomes of the approach depending on my assumptions. My first step is to define indicators of problem severity and discuss how I aggregate them to calculate those problems.

My ultimate goal is to quantify the severity of the problem and rank death and suffering now and in the future against each other. I named the basic index *ProbSev*, which is the global problem severity and is calculated based on death equivalents per year expressed as a proportion of global population. I consider these deaths in terms of current actual deaths, potential future deaths expressed as an annual rate, death equivalents to represent suffering, and death equivalents per year depending on future suffering.

I represent the estimated current global death rates for each particular problem as the *global deaths per year* (*GDPY*). This number has the most certainty. *Potential deaths per year* (*PDPY*) is the measure of risk of death from a future cause. I calculate the *PDPY* for a particular event from the total deaths estimated to result from a particular event multiplied by the probability that the event would occur in any particular year. Take, for example, the possibility that an asteroid will hit the Earth and cause extinction of all of humanity. As discussed in the second chapter, such an event is expected to occur once every half billion years. The annual probability is then 1/500,000,000 or 0.000000002 per year. This number multiplied times a population of 11 billion people leads to a *PDPY* of 2 deaths per year. The approach assigns the intrinsic value to the continuation of the human race, by using that as an upper bound of the index (assuming all people on Earth can only die in one category to avoid double counting).

As mentioned, I scale the index of potential deaths per year to the total population. This means that a problem that wiped out one fourth of the global population, as did the plague pandemic in the 1300s, when the population of the Earth was about half billion and about 125 million people died, would be weighted just as heavily as the same scale event happening now with 1.8 billion people dying.

© Springer Nature Switzerland AG 2019
W. Dodds, *The World's Worst Problems*,
https://doi.org/10.1007/978-3-030-30410-2

The index that makes suffering rates comparable to death rates is the *suffering death equivalents* per year (*SDEQ*). I calculate the *SDEQ* as the proportion of people on the planet suffering multiplied by a *suffering death multiplier* (*SDM*). It is an indicator of how many years of suffering are equivalent to one death. I will discuss relative rankings of suffering and how suffering scales against death in the following section.

Potential suffering death equivalents per year (*PSDEQ*) is an estimate of future suffering converted to death equivalents. This number is even less certain than the *SDEQ*. Nonetheless it will be calculated similarly to the *PDPY*. *PSDEQ* will be the probability that any event causes suffering per year, times the number of people suffering, and multiplied by suffering death multiplier. An example will illustrate this idea. Consider a somewhat smaller asteroid, this time one large enough to disrupt agriculture globally. As discussed previously, a regionally destructive collision with an asteroid with a diameter of 900 feet should occur roughly every 50,000 years. Let us assume this event causes half the people on Earth (3.5 billion people) to suffer for 5 years due to destabilization of agriculture and the global economy. Then the *PSDEQ* would be 5 ∗ 3.5 billion/50,000/20. The 20 is the *SDM* used to equate death to suffering. In this case, *PSDEQ* is 17,500 deaths per year per 7 billion people total. The number is considerably less than the roughly 6,000,000 childhood deaths per year related to hunger.

Many problems potentially cause both death and suffering. Therefore, for any individual problem, the problem severity in death equivalents per year is the sum of each of the proportions

$$PROBSEV = GDPY + PDPY + SDEQ + PSDEQ$$

Put into words, potential death equivalents each year proportional to total population is the sum of proportional global deaths per year, potential deaths per year, suffering death equivalents, and potential suffering death equivalents.

Given the wide range of the death and suffering associated with the various problems, I transform *PROBSEV* to a logarithmic scale. The log scale is similar to that of the Richter scale for earthquakes, so an increase by one number indicates a tenfold increase. The scale is forced to range from 0 to 10, with 0 indicating roughly one death per year and 10 indicating complete extinction of the human race. Specifically, the transformation is:

$$\text{Transformed } PROBSEV = \log_{10}\left(\left(PROBSEV + 1 * 10^{-10}\right) * 10^{10}\right)$$

References

Global Problems?

1. Vosoughi S, Roy D, Aral S (2018) The spread of true and false news online. Science 359(6380):1146–1151
2. Lazer DMJ, Baum MA, Benkler Y, Berinsky AJ, Greenhill KM, Menczer F et al (2018) The science of fake news. Science 359(6380):1094–1096. https://doi.org/10.1126/science.aao2998
3. Nyhan B, Reifler J (2010) When corrections fail: the persistence of political misperceptions. Polit Behav 32(2):303–330
4. Grinberg N, Joseph K, Friedland L, Swire-Thompson B, Lazer D (2019) Fake news on Twitter during the 2016 U.S. presidential election. Science 363(6425):374–378. https://doi.org/10.1126/science.aau2706
5. Nickerson RS (1998) Confirmation bias: a ubiquitous phenomenon in many guises. Rev Gen Psychol 2(2):175
6. Dodds WK (2008) Humanity's footprint: momentum, impact and our global environment. Columbia University Press, New York
7. Harris S (2011) The moral landscape: how science can determine human values. Simon and Schuster, New York
8. Pinker S (2018) Enlightenment now: the case for reason, science, humanism, and progress. Penguin, London
9. Xu X, Zuo X, Wang X, Han S (2009) Do you feel my pain? Racial group membership modulates empathic neural responses. J Neurosci 29(26):8525–8529
10. Lamm C, Decety J, Singer T (2011) Meta-analytic evidence for common and distinct neural networks associated with directly experienced pain and empathy for pain. NeuroImage 54(3):2492–2502
11. Kopnina H, Washington H, Gray J, Taylor B (2018) The "future of conservation" debate: defending ecocentrism and the Nature Needs Half movement. Biol Conserv 217:140–148
12. Leighton AH, Hughes CC (1955) Notes on Eskimo patterns of suicide. Southwest J Anthropol 11(4):327–338
13. Sepúlveda J, Murray C (2014) The state of global health in 2014. Science 345(6202):1275–1278. https://doi.org/10.1126/science.1257099

© Springer Nature Switzerland AG 2019
W. Dodds, *The World's Worst Problems*,
https://doi.org/10.1007/978-3-030-30410-2

Apocalypse

14. Olson P (2018) The spinning magnet: the electromagnetic force that created the modern world-and could destroy it. Nature Publishing Group, London
15. Williams MA, Ambrose SH, van der Kaars S, Ruehlemann C, Chattopadhyaya U, Pal J et al (2009) Environmental impact of the 73 ka Toba super-eruption in South Asia. Palaeogeogr Palaeoclimatol Palaeoecol 284(3–4):295–314
16. Mellars P, Gori KC, Carr M, Soares PA, Richards MB (2013) Genetic and archaeological perspectives on the initial modern human colonization of southern Asia. Proc Natl Acad Sci 110(26):10699–10704
17. Mason BG, Pyle DM, Oppenheimer C (2004) The size and frequency of the largest explosive eruptions on Earth. Bull Volcanol 66(8):735–748
18. Self S (2006) The effects and consequences of very large explosive volcanic eruptions. Philos Trans R Soc Lond A 364(1845):2073–2097
19. Berman B (2019) Earth-shattering. Little Brown and Company, New York
20. Piran T, Jimenez R (2014) Possible role of gamma ray bursts on life extinction in the universe. Phys Rev Lett 113(23):231102. https://doi.org/10.1103/PhysRevLett.113.231102
21. McLachlan JA, Arnold SF (1996) Environmental estrogens. Am Sci 84(5):452–461
22. Rees MJ (2003) Our final hour: a scientist's warning: how terror, error, and environmental disaster threaten humankind's future in this century--on earth and beyond. Basic Books, New York
23. Sofroniou A (2017) Philosophy of science and eschatology. Lulu.com

Disease Now and Potential Future Pandemics

24. Mee C (1990) How a mysterious disease laid low Europe's masses. Smithsonian 20(11):66–77
25. Diamond JM (1998) Guns, germs and steel: a short history of everybody for the last 13,000 years. Random House, London
26. Lindo J, Huerta-Sánchez E, Nakagome S, Rasmussen M, Petzelt B, Mitchell J et al (2016) A time transect of exomes from a Native American population before and after European contact. Nat Commun 7:13175. https://doi.org/10.1038/ncomms13175
27. Eyler JM (2001) The changing assessments of John Snow's and William Farr's cholera studies. Soz Präventivmed 46(4):225–232
28. Wang R, van Dorp L, Shaw LP, Bradley P, Wang Q, Wang X et al (2018) The global distribution and spread of the mobilized colistin resistance gene mcr-1. Nat Commun 9(1):1179. https://doi.org/10.1038/s41467-018-03205-z
29. Sepúlveda J, Murray C (2014) The state of global health in 2014. Science 345(6202):1275–1278. https://doi.org/10.1126/science.1257099
30. Wolfe N (2011) The viral storm: the dawn of a new pandemic age. Macmillan, London
31. Smith KF, Sax DF, Lafferty KD (2006) Evidence for the role of infectious disease in species extinction and endangerment. Conserv Biol 20(5):1349–1357
32. De Castro F, Bolker B (2005) Mechanisms of disease-induced extinction. Ecol Lett 8(1):117–126
33. Keesing F, Belden LK, Daszak P, Dobson A, Harvell CD, Holt RD et al (2010) Impacts of biodiversity on the emergence and transmission of infectious diseases. Nature 468(7324):647–652
34. Lipsitch M, Plotkin JB, Simonsen L, Bloom B (2012) Evolution, safety, and highly pathogenic influenza viruses. Science 336(6088):1529–1531
35. Henderson DA (1998) Bioterrorism as a public health threat. Emerg Infect Dis 4(3):488
36. Woodward A (2005) International regulations and agreements pertaining to bioterrorism. In: Encyclopedia of bioterrorism defense. Wiley, Hoboken

Hunger

37. Food and Agriculture Organization of the United Nations (FAOSTAT) (2011). http://faostat.fao.org/site/535/DesktopDefault.aspx. PageID

38. Gerland P, Raftery AE, Ševčíková H, Li N, Gu D, Spoorenberg T et al (2014) World population stabilization unlikely this century. Science 346(6206):234–237

39. Alexander P, Rounsevell MD, Dislich C, Dodson JR, Engström K, Moran D (2015) Drivers for global agricultural land use change: the nexus of diet, population, yield and bioenergy. Glob Environ Chang 35:138–147

40. Pauly D, Zeller D (2016) Catch reconstructions reveal that global marine fisheries catches are higher than reported and declining. Nat Commun 7:10244. https://doi.org/10.1038/ncomms10244

41. Driscoll CT, Mason RP, Chan HM, Jacob DJ, Pirrone N (2013) Mercury as a global pollutant: sources, pathways, and effects. Environ Sci Technol 47(10):4967–4983. https://doi.org/10.1021/es305071v

42. Crist E, Mora C, Engelman R (2017) The interaction of human population, food production, and biodiversity protection. Science 356(6335):260–264. https://doi.org/10.1126/science.aal2011

43. Fischer T, Byerlee D, Edmeades G (2012) Crop yields and food security: will yield increases continue to feed the world. In: Proceedings of the 12th Australian agronomy conference, pp 14–18. Australian Centre for International Agricultural Research (ACIAR), Bruce AU

44. Childers DL, Corman J, Edwards M, Elser JJ (2011) Sustainability challenges of phosphorus and food: solutions from closing the human phosphorus cycle. Bioscience 61(2):117–124

45. Gurian-Sherman D (2009) Failure to yield: evaluating the performance of genetically engineered crops. Union of Concerned Scientists, Cambridge, MA

46. Eisenhut M, Weber APM (2019) Improving crop yield. Science 363(6422):32–33. https://doi.org/10.1126/science.aav8979

47. Ray DK, Ramankutty N, Mueller ND, West PC, Foley JA (2012) Recent patterns of crop yield growth and stagnation. Nat Commun 3:1293

48. Mueller ND, Gerber JS, Johnston M, Ray DK, Ramankutty N, Foley JA (2012) Closing yield gaps through nutrient and water management. Nature 490(7419):254–257

49. WWAPU (2015) The United Nations world water development report 2015: water for a sustainable world. United Nations World Water Assessment Programme, Paris

50. Gleeson T, Wada Y, Bierkens MF, van Beek LP (2012) Water balance of global aquifers revealed by groundwater footprint. Nature 488(7410):197

51. Egan T (2006) The worst hard time: the untold story of those who survived the great American dust bowl. Houghton Mifflin Harcourt, Boston

52. Pimentel D, Harvey C, Resosudarmo P, Sinclair K, Kurz D, McNair M et al (1995) Environmental and economic costs of soil erosion and conservation benefits. Science 267(5201):1117–1123

53. D'Odorico P, Bhattachan A, Davis KF, Ravi S, Runyan CW (2013) Global desertification: drivers and feedbacks. Adv Water Resour 51:326–344

54. Collier P (2010) The plundered planet: why we must--and how we can--manage nature for global prosperity. Oxford University Press, New York

55. Godfray HCJ, Beddington JR, Crute IR, Haddad L, Lawrence D, Muir JF et al (2010) Food security: the challenge of feeding 9 billion people. Science 327(5967):812–818

Nuclear Weapons

56. Kristensen HM, Korda M (2019) United States nuclear forces, 2019. Bull At Sci 75(3):122–134. https://doi.org/10.1080/00963402.2019.1606503
57. SIPRI Institute (2019) SIPRI YEARBOOK 2019: armaments, disarmament and international security. Oxford University Press, Oxford
58. Mills MJ, Toon OB, Turco RP, Kinnison DE, Garcia RR (2008) Massive global ozone loss predicted following regional nuclear conflict. Proc Natl Acad Sci 105(14):5307–5312
59. Robock A, Oman L, Stenchikov GL (2007) Nuclear winter revisited with a modern climate model and current nuclear arsenals: still catastrophic consequences. J Geophys Res Atmos 112(D13):1–14
60. Robock A (2011) Nuclear winter is a real and present danger. Nature 473(7347):275

Global Environment in the Anthropocene

61. Hooke RL (2000) On the history of humans as geomorphic agents. Geology 28(9):843–846
62. Smil V (2013) Harvesting the biosphere: what we have taken from nature. MIT Press, Cambridge, MA
63. Smil V (2011) Harvesting the biosphere: the human impact. Popul Dev Rev 37(4):613–636
64. Bar-On YM, Phillips R, Milo R (2018) The biomass distribution on Earth. Proc Natl Acad Sci 115(25):6506–6511. https://doi.org/10.1073/pnas.1711842115
65. Jones KR, Venter O, Fuller RA, Allan JR, Maxwell SL, Negret PJ et al (2018) One-third of global protected land is under intense human pressure. Science 360(6390):788–791
66. Dodds WK (2008) Humanity's footprint: momentum, impact and our global environment. Columbia University Press, New York
67. Hoekstra AY, Mekonnen MM (2012) The water footprint of humanity. Proc Natl Acad Sci 109(9):3232–3237
68. Brown JH, Burnside WR, Davidson AD, DeLong JP, Dunn WC, Hamilton MJ et al (2011) Energetic limits to economic growth. Bioscience 61(1):19–26
69. Eitelberg DA, van Vliet J, Verburg PH (2015) A review of global potentially available cropland estimates and their consequences for model-based assessments. Glob Chang Biol 21(3):1236–1248
70. Ceballos G, Ehrlich PR, Barnosky AD, García A, Pringle RM, Palmer TM (2015) Accelerated modern human–induced species losses: entering the sixth mass extinction. Sci Adv 1(5):e1400253
71. Rockström J, Steffen W, Noone K, Persson Å, Chapin FS III, Lambin EF et al (2009) A safe operating space for humanity. Nature 461(7263):472
72. Pires M (2004) Watershed protection for a world city: the case of New York. Land Use Policy 21(2):161–175
73. Costanza R, D'Arge R, de Groot R, Farber S, Grasso M, Hannon B et al (1997) The value of the world's ecosystem services and natural capital. Nature 387:253–260
74. Dodds WK, Whiles MR (2010) Freshwater ecology: concepts and environmental applications of limnology, 2nd edn. Academic Press, Burlington
75. Arrow K, Cropper M, Gollier C, Groom B, Heal G, Newell R et al (2013) Determining benefits and costs for future generations. Science 341(6144):349–350. https://doi.org/10.1126/science.1235665
76. Cardinale BJ, Duffy JE, Gonzalez A, Hooper DU, Perrings C, Venail P et al (2012) Biodiversity loss and its impact on humanity. Nature 486(7401):59–67
77. Hull PM, Darroch SAF, Erwin DH (2015) Rarity in mass extinctions and the future of ecosystems. Nature 528(7582):345–351. https://doi.org/10.1038/nature16160

78. Costanza R, de Groot R, Sutton P, van der Ploeg S, Anderson SJ, Kubiszewski I et al (2014) Changes in the global value of ecosystem services. Glob Environ Chang 26:152–158

79. Brook EJ, Buizert C (2018) Antarctic and global climate history viewed from ice cores. Nature 558(7709):200

80. Miller JB, Lehman SJ, Montzka SA, Sweeney C, Miller BR, Karion A et al (2012) Linking emissions of fossil fuel CO_2 and other anthropogenic trace gases using atmospheric $^{14}CO_2$. J Geophys Res Atmos 117(D8):8302

81. Eisenhut M, Weber APM (2019) Improving crop yield. Science 363(6422):32–33. https://doi.org/10.1126/science.aav8979

82. Rohde R, Muller R, Jacobsen R, Perlmutter S, Rosenfeld A, Wurtele J et al (2013) Berkeley Earth temperature averaging process. Geoinfor Geostat Overview 1(2):13:20–100

83. Rohde R, Muller R, Jacobsen R, Muller E, Perlmutter S, Rosenfeld A et al (2013) A new estimate of the average Earth surface land temperature spanning 1753 to 2011. Geoinfor Geostat Overview 1(1):7:2

84. Tranter B, Booth K (2015) Scepticism in a changing climate: a cross-national study. Glob Environ Chang 33:154–164. https://doi.org/10.1016/j.gloenvcha.2015.05.003

85. Snyder CW (2016) Evolution of global temperature over the past two million years. Nature 538(7624):226–228. https://doi.org/10.1038/nature19798

86. Hsiang SM, Burke M, Miguel E (2013) Quantifying the influence of climate on human conflict. Science 341(6151):1235367. https://doi.org/10.1126/science.1235367

87. Phrampus BJ, Hornbach MJ (2012) Recent changes to the Gulf Stream causing widespread gas hydrate destabilization. Nature 490(7421):527–530

88. Spratt D, Dunlop I (2017) What lies beneath: the scientific understatement of climate risks breakthrough. National Centre for Climate Restoration, Melbourne, AU

89. Cheung WWL, Watson R, Pauly D (2013) Signature of ocean warming in global fisheries catch. Nature 497(7449):365–368. https://doi.org/10.1038/nature12156

90. Petrescu RV, Aversa R, Apicella A, Petrescu FI (2018) NASA sees first in 2018 the direct proof of ozone hole recovery. J Aircr Spacecr Technol 2(1):53–64

91. Barnett TP, Adam JC, Lettenmaier DP (2005) Potential impacts of a warming climate on water availability in snow-dominated regions. Nature 438(7066):303

92. Munia H, Guillaume J, Mirumachi N, Porkka M, Wada Y, Kummu M (2016) Water stress in global transboundary river basins: significance of upstream water use on downstream stress. Environ Res Lett 11(1):014002

93. Barnosky AD, Matzke N, Tomiya S, Wogan GOU, Swartz B, Quental TB et al (2011) Has the Earth's sixth mass extinction already arrived? Nature 471(7336):51–57

94. Kroeker KJ, Kordas RL, Crim RN, Singh GG (2010) Meta-analysis reveals negative yet variable effects of ocean acidification on marine organisms. Ecol Lett 13(11):1419–1434. https://doi.org/10.1111/j.1461-0248.2010.01518.x

95. McCauley DJ, Pinsky ML, Palumbi SR, Estes JA, Joyce FH, Warner RR (2015) Marine defaunation: animal loss in the global ocean. Science 347(6219):1255641. https://doi.org/10.1126/science.1255641

96. Estes JA, Terborgh J, Brashares JS, Power ME, Berger J, Bond WJ et al (2011) Trophic downgrading of planet Earth. Science 333(6040):301–306

97. Reid AJ, Carlson AK, Creed IF, Eliason EJ, Gell PA, Johnson PT et al (2018) Emerging threats and persistent conservation challenges for freshwater biodiversity. Biol Rev 94:849–873

98. MacDougall AS, McCann KS, Gellner G, Turkington R (2013) Diversity loss with persistent human disturbance increases vulnerability to ecosystem collapse. Nature 494(7435):86–89

99. Goulson D, Nicholls E, Botías C, Rotheray EL (2015) Bee declines driven by combined stress from parasites, pesticides, and lack of flowers. Science 347(6229):1255957

100. Garibaldi LA, Steffan-Dewenter I, Winfree R, Aizen MA, Bommarco R, Cunningham SA et al (2013) Wild pollinators enhance fruit set of crops regardless of honey bee abundance. Science 339(6127):1608–1611

101. Eyre BD, Cyronak T, Drupp P, De Carlo EH, Sachs JP, Andersson AJ (2018) Coral reefs will transition to net dissolving before end of century. Science 359(6378):908–911. https://doi.org/10.1126/science.aao1118
102. Bumb BL, Baanante CA (1996) World trends in fertilizer use and projections to 2020. IFPRI, Washington, DC
103. Dodds WK, Bouska WW, Eitzmann JL, Pilger TJ, Pitts KL, Riley AJ et al (2009) Eutrophication of US freshwaters: analysis of potential economic damages. Environ Sci Technol 43(1):12–19. https://doi.org/10.1021/Es801217q
104. Rabotyagov S, Kling C, Gassman P, Rabalais N, Turner R (2014) The economics of dead zones: causes, impacts, policy challenges, and a model of the Gulf of Mexico hypoxic zone. Rev Environ Econ Policy 8(1):58–79
105. Diaz RJ, Rosenberg R (2008) Spreading dead zones and consequences for marine ecosystems. Science 321(5891):926–929
106. Stehle S, Schulz R (2015) Agricultural insecticides threaten surface waters at the global scale. Proc Natl Acad Sci 112(18):5750–5755. https://doi.org/10.1073/pnas.1500232112
107. Li J, Cao J, Zhu Y-G, Chen Q-L, Shen F, Wu Y et al (2018) Global survey of antibiotic resistance genes in air. Environ Sci Technol 52(19):10975–10984. https://doi.org/10.1021/acs.est.8b02204

By the Numbers: Ranking the Problem

108. Gerland P, Raftery AE, Ševčíková H, Li N, Gu D, Spoorenberg T et al (2014) World population stabilization unlikely this century. Science 346(6206):234–237
109. You D, New JR, Wardlaw T (2012) Levels and trends in child mortality. Report 2012. Estimates developed by the UN Inter-agency Group for Child Mortality Estimation
110. Stop_the_Hunger (2019). http://www.stopthehunger.com/
111. World Health Organization (WHO) (2011) World Health Statistics 2011. World Health Organization, Geneva. World Health Statistics 2012
112. FAO, UNICEF, WFP and WHO (2017) The state of food security and nutrition in the world 2017: building resilience for peace and food security. Food and Agriculture Organization of the United Nations (FAO), Rome
113. Rees MJ (2003) Our final hour: a scientist's warning: how terror, error, and environmental disaster threaten humankind's future in this century--on earth and beyond. Basic Books, New York
114. Baum S, de Neufville R, Barrett A (2018) A model for the probability of nuclear war. Global Catastrophic Risk Institute Working Paper:18-1
115. Barnosky AD, Hadly EA, Bascompte J, Berlow EL, Brown JH, Fortelius M et al (2012) Approaching a state shift in Earth's biosphere. Nature 486(7401):52–58
116. Mora C, Dousset B, Caldwell IR, Powell FE, Geronimo RC, Bielecki CR et al (2017) Global risk of deadly heat. Nat Clim Chang 7(7):501
117. Chapman CR, Morrison D (1994) Impacts on the Earth by asteroids and comets: assessing the hazard. Nature 367(6458):33
118. Board SS, Council NR (2010) Defending planet Earth: near-Earth-object surveys and hazard mitigation strategies. National Academies Press, Washington, DC
119. Nowaczyk N, Arz H, Frank U, Kind J, Plessen B (2012) Dynamics of the Laschamp geomagnetic excursion from Black Sea sediments. Earth Planet Sci Lett 351:54–69
120. Self S (2006) The effects and consequences of very large explosive volcanic eruptions. Philos Trans R Soc Lond A 364(1845):2073–2097
121. Piran T, Jimenez R (2014) Possible role of gamma ray bursts on life extinction in the universe. Phys Rev Lett 113(23):231102. https://doi.org/10.1103/PhysRevLett.113.231102

122. Global_Health_Observatory (2018). http://apps.who.int/gho/data/node.home. Accessed 5 Nov 2018

123. Iuliano AD, Roguski KM, Chang HH, Muscatello DJ, Palekar R, Tempia S et al (2018) Estimates of global seasonal influenza-associated respiratory mortality: a modelling study. Lancet 391(10127):1285–1300. https://doi.org/10.1016/S0140-6736(17)33293-2

124. Institute_for_Health_Metrics_and_Evaluation (2016) Global burden of disease study 2016 (GBD 2016) data resources. http://ghdx.healthdata.org/gbd-2016. Accessed 6 Nov 2018.

125. Wolfe N (2011) The viral storm: the dawn of a new pandemic age. Macmillan, London

126. Niall PA, Johnson S, Mueller J (2002) Updating the accounts: global mortality of the 1918–1920 "Spanish" influenza pandemic. Bull Hist Med 76(1):105–115

What Do Other People Think the Worst Problems in the World Are?

127. Diamond J (2019) Upheaval: turning points for nations in crisis. Little, Brown and Company, New York

Progress Toward Solving the Problems and Potential Costs of Solutions

128. Brown JH, Burnside WR, Davidson AD, Delong JR, Dunn WC, Hamilton MJ et al (2011) Energetic limits to economic growth. Bioscience 61(1):19–26

129. Jean N, Burke M, Xie M, Davis WM, Lobell DB, Ermon S (2016) Combining satellite imagery and machine learning to predict poverty. Science 353(6301):790–794. https://doi.org/10.1126/science.aaf7894

130. Cree A, Kay A, Steward J (2012) The economic and social cost of illiteracy: a snapshot of illiteracy in a global context. World Literacy Foundation, Melbourne

131. McCarthy DP, Donald PF, Scharlemann JPW, Buchanan GM, Balmford A, Green JMH et al (2012) Financial costs of meeting global biodiversity conservation targets: current spending and unmet needs. Science 338(6109):946–949. https://doi.org/10.1126/science.1229803

132. Wilting HC, Schipper AM, Bakkenes M, Meijer JR, Huijbregts MAJ (2017) Quantifying biodiversity losses due to human consumption: a global-scale footprint analysis. Environ Sci Technol 51(6):3298–3306. https://doi.org/10.1021/acs.est.6b05296

133. Malik A, Lan J, Lenzen M (2016) Trends in global greenhouse gas emissions from 1990 to 2010. Environ Sci Technol 50(9):4722–4730. https://doi.org/10.1021/acs.est.5b06162

134. Boyd JW, Bagstad KJ, Ingram JC, Shapiro CD, Adkins JE, Casey CF et al (2018) The natural capital accounting opportunity: let's really do the numbers. Bioscience 68(12):940–943

135. Thompson KM, Tebbens RJD (2007) Eradication versus control for poliomyelitis: an economic analysis. Lancet 369(9570):1363–1371

136. UNICEF (2015) Committing to child survival: a promise renewed. eSocialSciences

137. World Health Organization (2018) Europe observes a 4-fold increase in measles cases in 2017 compared to previous year, vol 6. World Health Organization, Copenhagen, Denmark

138. Board SS, Council NR (2010) Defending planet Earth: near-Earth-object surveys and hazard mitigation strategies. National Academies Press, Washington, DC

139. Cartier KMS (2018) Are we prepared for an asteroid headed straight to Earth? Eos 99. https://doi.org/10.1029/2018EO101939

140. Baum SD, Denkenberger DC, Pearce JM, Robock A, Winkler R (2015) Resilience to global food supply catastrophes. Environ Syst Decis 35(2):301–313. https://doi.org/10.1007/s10669-015-9549-2

141. Willett S (2003) Costs of disarmament--disarming the costs: nuclear arms control and nuclear rearmament. United Nations Publications UNIDIR, Geneva

Technochimp: An African Savannah Survivor Looking for Solutions in the Modern World

142. Diamond J (2014) The third chimpanzee. Oneworld Publications, London
143. Hein G, Morishima Y, Leiberg S, Sul S, Fehr E (2016) The brain's functional network architecture reveals human motives. Science 351(6277):1074–1078. https://doi.org/10.1126/science.aac7992
144. Hofman JM, Sharma A, Watts DJ (2017) Prediction and explanation in social systems. Science 355(6324):486–488. https://doi.org/10.1126/science.aal3856
145. Zuk M (2013) Paleofantasy: what evolution really tells us about sex, diet, and how we live. WW Norton & Company, New York
146. Falk A, Szech N (2013) Morals and markets. Science 340(6133):707–711
147. Henrich J, McElreath R, Barr A, Ensminger J, Barrett C, Bolyanatz A et al (2006) Costly punishment across human societies. Science 312(5781):1767–1770
148. Levin SA (2014) Public goods in relation to competition, cooperation, and spite. Proc Natl Acad Sci 111(Supplement 3):10838–10845. https://doi.org/10.1073/pnas.1400830111
149. Sigmund K, De Silva H, Traulsen A, Hauert C (2010) Social learning promotes institutions for governing the commons. Nature 466(7308):861–863
150. Galbiati R, Henry E, Jacquemet N (2018) Dynamic effects of enforcement on cooperation. Proc Natl Acad Sci 115(49):12425–12428. https://doi.org/10.1073/pnas.1813502115
151. Kruger J, Dunning D (1999) Unskilled and unaware of it: how difficulties in recognizing one's own incompetence lead to inflated self-assessments. J Pers Soc Psychol 77(6):1121
152. Lord CG, Ross L, Lepper MR (1979) Biased assimilation and attitude polarization: the effects of prior theories on subsequently considered evidence. J Pers Soc Psychol 37(11):2098
153. Nickerson RS (1998) Confirmation bias: a ubiquitous phenomenon in many guises. Rev Gen Psychol 2(2):175
154. Bodin Ö (2017) Collaborative environmental governance: achieving collective action in social-ecological systems. Science 357(6352):eaan1114. https://doi.org/10.1126/science.aan1114
155. Lloyd WF (1980) WF Lloyd on the checks to population. Popul Dev Rev 6(3):473–496
156. Boyd R, Richerson PJ, Meinzen-Dick R, De Moor T, Jackson MO, Gjerde KM et al (2018) Tragedy revisited. Science 362(6420):1236–1241. https://doi.org/10.1126/science.aaw0911
157. Rustagi D, Engel S, Kosfeld M (2010) Conditional cooperation and costly monitoring explain success in forest commons management. Science 330(6006):961–965. https://doi.org/10.1126/science.1193649
158. Phillips T (2015) Human altruism and cooperation explainable as adaptations to past environments no longer fully evident in the modern world. Q Rev Biol 90(3):295–314. https://doi.org/10.1086/682589

International Cooperation

159. Gat A (2013) Is war declining–and why? J Peace Res 50(2):149–157
160. Whitman J (2005) Limits of global governance. Routledge, London
161. Diamond J (2019) Upheaval: turning points for nations in crisis. Little, Brown and Company, New York
162. Snidal D (1991) Relative gains and the pattern of international cooperation. Am Polit Sci Rev 85(3):701–726
163. Collier P (2010) The plundered planet: why we must--and how we can--manage nature for global prosperity. Oxford University Press, New York
164. Pinker S (2018) Enlightenment now: the case for reason, science, humanism, and progress. Penguin, London
165. Rigby M, Park S, Saito T, Western LM, Redington AL, Fang X et al (2019) Increase in CFC-11 emissions from eastern China based on atmospheric observations. Nature 569(7757):546–550. https://doi.org/10.1038/s41586-019-1193-4
166. Daniel A, Guardans R, Harner T (2018) The contribution of environmental monitoring to the review of the effectiveness of environmental treaties. Environ Sci Technol 52(1):1–2. https://doi.org/10.1021/acs.est.7b06148
167. Nordhaus WD (2013) The climate casino: risk, uncertainty, and economics for a warming world. Yale University Press, New Haven
168. Blair BG (2011) The logic of accidental nuclear war. Brookings Institution Press, Washington, DC
169. Aginam O (2002) International law and communicable diseases. Bull World Health Organ 80(12):946–951

Consilience, Global Socioeconomic Political Enlightenment, and Socioenvironmental Restoration

170. Wilson EO (1999) Consilience: the unity of knowledge, vol 31. Vintage, New York
171. Kasper EK, Gerstenblith G, Hefter G, Van Anden E, Brinker JA, Thiemann DR et al (2002) A randomized trial of the efficacy of multidisciplinary care in heart failure outpatients at high risk of hospital readmission. J Am Coll Cardiol 39(3):471–480
172. Turchin P, Currie TE, Whitehouse H, François P, Feeney K, Mullins D et al (2018) Quantitative historical analysis uncovers a single dimension of complexity that structures global variation in human social organization. Proc Natl Acad Sci 115(2):E144–E151. https://doi.org/10.1073/pnas.1708800115
173. Liu J, Mooney H, Hull V, Davis SJ, Gaskell J, Hertel T et al (2015) Systems integration for global sustainability. Science 347(6225):1258832. https://doi.org/10.1126/science.1258832
174. Liu Y-Y, Slotine J-J, Barabasi A-L (2011) Controllability of complex networks. Nature 473(7346):167–173
175. Subrahmanian VS, Kumar S (2017) Predicting human behavior: the next frontiers. Science 355(6324):489–489. https://doi.org/10.1126/science.aam7032
176. Hackett CP, Grim BJ (2012) The global religious landscape: a report on the size and distribution of the world's major religious groups as of 2010. Pew Research Center, Pew Forum on Religion & Public Life, Washington, DC
177. Dasgupta P (2002) Is contemporary economic development sustainable? Ambio 31(4):269–271
178. Dasgupta PS, Ehrlich PR (2013) Pervasive externalities at the population, consumption, and environment nexus. Science 340(6130):324–328

179. Cumming GS, von Cramon-Taubadel S (2018) Linking economic growth pathways and environmental sustainability by understanding development as alternate social–ecological regimes. Proc Natl Acad Sci 115(38):9533–9538. https://doi.org/10.1073/pnas.1807026115

180. Hauser OP, Rand DG, Peysakhovich A, Nowak MA (2014) Cooperating with the future. Nature 511(7508):220–223. https://doi.org/10.1038/nature13530

181. Wike R, Simmons K, Stokes B, Fetterolf J (2017) Globally, broad support for representative and direct democracy. Pew Research Center, Washington, DC

182. Pinker S (2018) Enlightenment now: the case for reason, science, humanism, and progress. Penguin, London

183. Scheffer M, van Bavel B, van de Leemput IA, van Nes EH (2017) Inequality in nature and society. Proc Natl Acad Sci 114(50):13154–13157. https://doi.org/10.1073/pnas.1706412114

184. Rifkin J (2011) The third industrial revolution: how lateral power is transforming energy, the economy, and the world. Macmillan, New York

185. Dietz T, Ostrom E, Stern PC (2003) The struggle to govern the commons. Science 302:1907–1912

186. Centola D, Becker J, Brackbill D, Baronchelli A (2018) Experimental evidence for tipping points in social convention. Science 360(6393):1116–1119

187. Nyborg K, Anderies JM, Dannenberg A, Lindahl T, Schill C, Schlüter M et al (2016) Social norms as solutions. Science 354(6308):42–43. https://doi.org/10.1126/science.aaf8317

188. Aral S, Walker D (2012) Identifying influential and susceptible members of social networks. Science 337(6092):337–341. https://doi.org/10.1126/science.1215842

189. Woolley AW, Chabris CF, Pentland A, Hashmi N, Malone TW (2010) Evidence for a collective intelligence factor in the performance of human groups. Science 330(6004):686–688. https://doi.org/10.1126/science.1193147

190. Mellers B, Stone E, Murray T, Minster A, Rohrbaugh N, Bishop M et al (2015) Identifying and cultivating superforecasters as a method of improving probabilistic predictions. Perspect Psychol Sci 10(3):267–281. https://doi.org/10.1177/1745691615577794

191. Zlatev JJ, Daniels DP, Kim H, Neale MA (2017) Default neglect in attempts at social influence. Proc Natl Acad Sci 114(52):13643–13648. https://doi.org/10.1073/pnas.1712757114

Index

© Springer Nature Switzerland AG 2019
W. Dodds, *The World's Worst Problems*,
https://doi.org/10.1007/978-3-030-30410-2